Industrial Ecology

Series Editor

Syed Abdul Rehman Khan, Tsinghua University, Beijing, Beijing, China

Industrial ecology and circular economy is a peer-reviewed book series that focuses on different disciplinary approaches to waste management, sustainable practices & strategies on different scientific, societal, pyschological, technological, economic, governance, and cultural and political aspects of the ongoing and emerging debate. This primary goal of this series is to offer scientists from different school of thoughts and institutions a platform for scientific analysis and debate.

Undeniably, Industrial ecology is a rapidly growing field that systematically examines local, regional and global materials and energy uses and flows in products, processes, industrial sectors and economies. It focuses on the potential role of industry in reducing environmental burdens throughout the product life cycle from the extraction of raw materials, to the production of goods, to the use of those goods and to the management of the resulting wastes. Industrial ecology is ecological in that it (1) places human activity—industry in the very broadest sense—in the larger context of the biophysical environment from which we obtain resources and into which we place our wastes, and (2) looks to the natural world for models of highly efficient use of resources, energy and byproducts. By selectively applying these models, the environmental performance of industry can be improved. Industrial ecology sees corporate entities as key players in the protection of the environment, particularly where technological innovation is an avenue for environmental improvement. As repositories of technological expertise in our society, corporations provide crucial leverage in attacking environmental problems through product and process design.

More information about this series at http://www.springer.com/series/16605

Charbel Jose Chiappetta Jabbour ·
Syed Abdul Rehman Khan
Editors

Sustainable Production and Consumption Systems

Springer

Editors
Charbel Jose Chiappetta Jabbour
Emlyon Business School
Ecully, France

Lincoln International Business School
Lincoln, UK

Syed Abdul Rehman Khan
School of Engineering and Management
Xuzhou University of Technology
Xuzhou, Jiangsu Province, China

Beijing Key Laboratory of Urban Spatial
Information Engineering
Beijing, China

ISSN 2730-5775 ISSN 2730-5783 (electronic)
Industrial Ecology
ISBN 978-981-16-4759-8 ISBN 978-981-16-4760-4 (eBook)
https://doi.org/10.1007/978-981-16-4760-4

This Springer imprint is published by the registered company Springer Nature Singapore Pte Ltd.
The registered company address is: 152 Beach Road, #21-01/04 Gateway East, Singapore 189721, Singapore

Contents

Assessing EU's Progress and Performance with Regard to SDG-12 Targets and Indicators

Jean-Vasile Andrei, Marius Constantin, and Ignacio de los Ríos Carmenado

Abstract Finding a pathway between high levels of economic competitiveness and sustainability is challenging. The global economy has been long-calling for a paradigm shift as far as the consumption and production models have been concerned over the last decades. In this regard, transforming our world to the imperative vision described in the 2030 Agenda for Sustainable Development can be made possible if global markets respond efficiently to the goals and targets set in the 2030 Agenda. Taking this into consideration, business models, production patterns, purchasing and consumption habits represent key elements that are connected to the indictors designed to monitor the progress and performance achieved with regard to the SDG 12: tracking resource management efficiency, the circularity of economic activities, production and consumption footprint and others. Given the importance of quantifying results with the aim of ensuring the timely achievement of the SDG KPIs, the objective of this research was to carry out a synoptic quantitative analysis on EU-27's progress and performance with regard to the targets and indicators of SDG 12—not only from the perspective of the well-defined methodological framework described in Agenda, but also from the perspective of the scientific results published in the papers specific to the field of sustainable production and consumption patterns. Findings highlight the ongoing ardent transition to a more resource-efficient and circular European economic system. In the race for shifting the production and consumption patterns from the conventional model to the one that has the sustainability factor at its core in the EU-27, the vectors of change proved to be Finland, Austria and Latvia.

J.-V. Andrei
Petroleum-Gas University of Ploiesti, Ploiesti, Romania

National Institute for Economic Research 'Costin C. Kiritescu', Romanian Academy, 050711 Bucharest, Romania

M. Constantin (✉)
Bucharest University of Economic Studies, Bucharest, Romania
e-mail: marius.constantin@eam.ase.ro

I. de los Ríos Carmenado
Universidad Politécnica De Madrid, Madrid, Spain

C. J. Chiappetta Jabbour and S. A. R. Khan (eds.), *Sustainable Production and Consumption Systems*, Industrial Ecology,
https://doi.org/10.1007/978-981-16-4760-4_1

Keywords Sustainable consumption · Sustainable production · European green deal · Impact assessment

1 Introduction

Production and consumption represent the main factors that drive the global economy, with direct impact on the natural capital. Despite the fact that the economy is driven by such a complex and dynamyc system with social and environmental implications, giving importance only the economic aspect of the global market is no longer sufficient, it probably never had been, but especially now, at the beginning of the twenty-twenties, when the efficient management of the natural resources is more important than ever.

Humankind is facing multifaceted socio-economic and environmental issues, which are threatening the quality of life of future generations. The unsustainable and irrational use of natural resources has been causing planetary health havoc and has determined decision makers to find intelligent, cost-efficient and sustainable solutions to such issues. In a global context characterized by the fact that the global material footprint cannot keep up with the constant population growth and economic output, the United Nations have elaborated a strategy designed to provide answers through a coherent long-term action plan aiming to solve such ardernt problems in a sustainable manner. This strategy is called the 2030 Agenda for Sustainable Development (United Nations 2015). Its scope goes beyong the topics of sustainable consumption and production patterns—it tackles a variety of global challenges that humankind has been facing for decades now, with a special focus on eradicating poverty in any form or dimension. Its general aim is to ensure and consolidate high levels of well-being and economic prosperity, not only for current generations, but also for those to come, in such a manner that the environment is protected at all times.

Although the 2030 Agenda for Sustainable Development provides a holistic view on multidimensional issues and references 17 sustainable development goals (SDGs) and 169 specific targets, this book chapter is particularly focused only on SDG 12 and its targets, specific to responsible consumption and production patterns. Of course, there are certain dependencies among the SDGs, especially if considering their interactions (Pradhan et al. 2017), in some other cases, studies have proven SDGs synergy (van Noordwijk et al. 2018; Fader et al. 2018), yet studying these aspects does not represent the objective of this research.

This study aims at assessing EU-27's progress and performance with regard to the targets and indicators of the 2030 Agenda for Sustainable Development's Goal 12—"*Ensure sustainable consumption and production patterns*" from the perspective of the well-defined methodological framework described in Agenda, as well as from the perspective of the scientific results published in the papers specific to the field of sustainable production and consumption patterns. The objective of this research was reached by carrying out a standard statistical analysis at the level of EU-27 members on the one hand, and, on the other hand, the same research objective was met by applying another quantitative research method to a larger set of metadata: the bibliometric analysis, which was carried out based on the Web of Science-extracted metadata of the identified indexed scientific publications concerning the topic of sustainable consumption and production patterns.

Since implementing the 2030 Agenda for Sustainable Development is an incredibly complex process demanding global collective effort, it also demands for rigorous monitoring and constant evaluation of success in relation to the strategy's objectives. However, despite the fact that the literature contains numerous studies in which systematic analyses were carried out to explore the degree of successful SDGs implementation (Allen et al. 2018; Diaz-Sarachaga et al. 2018; Terama et al. 2016), the literature contains very few studies related to the goal 12 specifically.

This research complements the existing literature by approaching the topic of EU-27's progress with regard to the targets and indicators specific to sustainable consumption and production in a 'double-edged quantitative research manner' and an assessment method was proposed. Not only is the progress approached through the lens of the specific indicator framework of the 2030 Agenda for Sustainable Development, but progress is also approached according to the scientific performance of the research publications elaborated on this topic.

This book chapter is structured in four sections. The first section, *literature review*, contains a summary of relevant empirical studies elaborated on the topic of sustainable consumption and production patterns. The research methodology section grounds the methodological background for the third the section, which was designed for commenting and discussing the research findings. The latter points out the ongoing ardent transition to a more resource-efficient and circular economic system in the EU-27, while focusing on the progress made by each state with respect to SDG 12, based on the 2030 Agenda's framework and on the scientific interest looked upon through the lens of intensity of the documents published and indexed in the Web of Science database. The last section of this book chapter was dedicated to summarize the main findings of the research. At the end of this section, the limitations of this study were explained and further research directions were proposed.

2 Literature Review

The growth in population and the intensification of resource usage has direct and significant impact over the last few decades on the deterioration of the natural capital. It has become clear that the future of consumption and production patterns is dependent on the to transition to a more sustainable path, aiming to protect natural resources as much as possible, as well as efficiently manage them.

Bocken et al. (2014) identified the three main vectors that best-define the industrial sustainability agenda: eco-innovations, CSR (corporate social responsibility) and ecological efficiency. Their vision is following the model described by Evans et al. (2009): the previously mentioned vector contribute to reducing energy consumption, as well as resource intensity, and per-unit emissions and waste. Moreover, the research of Bocken et al. is grounded on Jackson and Senker's study (2011). In their article, the two authors described the features of the route to a world characterized by sustainable consumption and production patterns: (a) a system functioning according to the principle of consumption minimization or cap-based restriction; (b) a system that has properly functions in harmony with maximizing socio-economic and environmental benefits, instead of prioritizing only the economic growth factor; (c) a system enhancing creativity and innovation that provides the instruments for fulfilling work experiences in a sustainable and efficient manner; (d) a coherent and circular system articulated around the principles of functionality and experience delivery, instead of ownership (sharing economy)—therefore, a system built on collaboration, instead of exacerbated competition.

As far as the sharing economy and collaborative consumption models are concerned, Hamari et al. (2016) consider that, through the usage of information and communication technologies, humankind can change the consumption paradigm and ensure more inclusive and sustainable patterns, simultaniously with the mitigation socio-economic and environmental issues: pollution, hyper-consumption, poverty and others. Not only that, but their research also highlighted that the collaborative consumption might have some roots dealing with attitude-behavior gaps, since some people are still reticent to the concept of sharing economy and their verbally-expressed positive opinion on the model does not find its translation into actual practical actions.

Dinu et al. (2020b) argue for the importance of empowering sustainable production behaviors and consumption patterns in the context of ensuring a more cleaner and wealthier future for generations to come through different methods: (a) by reconsidering the necessity of each economic activity carried out and trying finding ways to decrease of the amount of greenhouse gas emissions generated; (b) focus on recycling and finding ways to increase energy productivity; (c) proper management of waste, especially in the context of the imperative transition to the circular economy.

Sustainable consumption and production need to be respected in the agricultural sector as well. Papargyropoulou et al. (2014) proved that food waste has significant global socio-economic and environmental implications and argued for the importance of preventing food waste and reducing food surplus through the efficient management of the food supply chain. In this regard, the depletion of natural resources would be mitigated by tackling food suplus and food waste holistically, throughout the entire global food supply chain, not only at the level of food consumption, but also as far as production is concerned. This is necessary because the agricultural sector puts a lot of pressure on the natural capital (Mekonnen and Hoekstra 2012) and generates considerable amounts of greenhouse gas emissions, at the level of all the activities carried out to ensure global food security. Constantin et al. (2021) studied the relation between agricultural productivity and its impact on the generation of greenhouse gas emisision in the EU-27, arguing that, despite its important contribution to ensuring food security (Istudor et al. 2019), agriculture needs to adapt to the exigencies of the new Common Agricultural Policy that calls for this sector to become more 'green' (environmentally-friendly) and circular (Drăgoi et al. 2018). This transition is encouraged and financed through specific European Funds, such as The European

Agricultural Fund for Rural Development and the European Agricultural Guarantee Fund (Dinu et al. 2020a).

'Green' consumption might appear as an oxymoron, since, on the one hand, it implies the conservation of resources, and, on the other hand, the nature of consumption itself refers to the 'destruction' of resources. The key takeaway is that 'green' consumption is focused on the rational consumer choice, stressing the role of attitudes, social norms, personal values and morality in the transition to healthy and sustainable lifestyles (Peattie 2010). In their research, Thøgersen and Ölander (2002) also argue that sustainable consumption patterns are influenced by personal values/priorities of the individual.

Røpke (2009) stressed on the immense global challenge of promoting and actually transitioning to the sustainable consumption model, especially in the context of constant real income increases, which, in the end, enables higher levels of production and consumption. In this regard, as far as the COVID-19 pandemic is concerned, in some cases, this sanitary, socio-economic crisis acted as a step back in the face of many emerging entrepreneurs and represented a pause in the intensification of economic activities (Ignat and Constantin 2020a), with the exception of some sectors: the manufacturing sector, especially in the case of sanitary products (Sarkodie and Owusu 2020; Selam 2020; Aragaw 2020), the agricultural and transportation sectors (Bracale and Vaccaro 2020; Ignat and Constantin 2020b; Rowan and Laffey 2020; Gray 2020).

Peattie and Peattie's study (2009) provides a perspective on how the discipline of social marketing can contribute to improve consumption patterns in the direction of its reduction, highlighting the role of promoting sustainable lifestyles through health-oriented social marketing campaigns. For social marketers, contributing to the transition to a more sustainable economy and society in a just manner is challenging—they have to maintain an equilibrium between lower consumer expectations, as well as delivering the best possible product/service quality to them. Yet, this form of marketing relies on the extended responsibility policy of the producers: both financially and physically, even after the lifetime of their products (Lebel and Lorek 2008), through recycling (Ignat and Constantin 2021).

Increasing productivity through heavy work investment on performant technologies and greater quantities is not always the solution for a healthy and sustainable lifestyle (Pătărlăgeanu et al. 2020a, b). Creating both competitive (profitable) and useful options for the market requires the joint work of producers and consumers to assess and design the specific needs of each in a sustainable manner. This collaboration is very much needed in the context of the Just Transition (Voicu-Dorobanțu et al. 2021) and of the implementation of the European Green Deal (Volintiru et al. 2019).

Although the literature is rich on papers approaching the topics of sustainable consumption and production, not much was written on evaluating the progress made with respect to SDG 12. In the early studies of Narayanaswamy and Stone (2007); Mont and Plepys (2008), these researcher identified the knowledge gap in this regard. A decade later, Roy and Singh (2017) elaborated a systematic literature review on the literature specific to the topics of sustainable consumption and production, mapping the progress made in terms of scientific work. Probably the most recent and relevant paper elaborated on the topic of this book chapter is that written by Gasper et al. (2019). They examined the processes of formulation of the Goal 12 and showed that the targets and indicators specific to SDG 12 reflect both practical/technical considerations and the political process in translation of such a broad concept, sustainable consumption and production, into the SDG format. Not only that, but the authors cosider that the indicators designed to measure progress on the SDG 12 targets does the contrary—they narrow its scope and ambition.

3 Research Methodology

The assessment on EU-27's progress and performance with regard to SDG 12 implies the need for a quantitative research method—a statistical analysis. In accordance with the objective of this research, a performance framework was designed based on the data corresponding to five relevant indicators, which were gathered from the Eurostat open-access database in April 2021: (1) resource productivity and domestic material consumption; (2) circular material use rate; (3 and 4) the nominal generation of hazardous and non-hazardous waste, excluding major mineral waste; (5) gross value added in environmental goods and services sector, as described in Table 1.

The first indicator included in Table 1 presents the GDP divided by domestic material consumption. The domestic material consumption refers to the amount of all materials running in the economy. This indicator is used to monitor EU-27's progress towards SDG 12, as well as SDG 8—on decent work and economic growth.

The second indicator, the circular material use rate, measures the share of material recovered and reintegrated into the economy, therefore saving extraction of primary

Table 1 The indicators used to assess EU's progress and performance with regard to SDG-12

No	Indicator name	Online data code	Unit of measure
1	Resource productivity and domestic material consumption	SDG_12_20	Euro per kilogram, chain linked volumes (2015)
2	Circular material use rate	SDG_12_41	Percentage
3	Generation of hazardous waste, excluding major mineral wastes	SDG_12_50	Kilograms per capita
4	Generation of non-hazardous waste, Excluding major mineral waste	SDG_12_50	Kilograms per capita
5	Gross value added in environmental goods and services sector	SDG_12_61	Percetange of gross domestic product (GDP)

Source Authors' selection of indicators from the Eurostat database (2021)

raw materials. This indicator can be refered to as the circularity rate, since it represents the ratio of circular use of materials in the economy to the overall used material. Higher circularity rate indicates the patterns of sustainable consumption and production due to more secondary materials being substitute for prirary raw materials.

The third indicator quantifies the nominal hazardous waste generated throughout a year, while the fourth indicator quantifies the nominal non-hazardous waste. Major mineral wastes, dredging spoils and soil were excluded.

The fifth indicator deals with the environmental goods and services sector, which is that part of a country's economy engaged in producing goods and services designed for environmental protection activities, as well as for the proper management of resources. The gross value added represents only the contribution of this sector to GDP, definied as the difference between the value of environmental goods and services sector's output and intermediate consumption.

4 Methodology of Bibliometric Analysis—Assessment on Scientific Work

The second part of this quantitative study is based on the method of bibliometric analysis, specific to scientometrics. Scientometrics and bibliometrics have grown in popularity for assessing the productivity, growth and interrelationships between scientific publications (Hood and Wilson 2001). Considering the objective of this research, this quantitative method represents a viable tool for assessing the scientific progress towards SDG 12, as far the production and quality of scientific resources are concerned.

Developing the bibliometric analysis in this book chapter relied on applying the VOSviewer procedures, as designed by in the work of van Eck and Waltman (2010, 2019). Using the 1.16.16 version of this software tool, VOSviewer facilitated the creations of maps, based on network metadata, as well as the visualizing and exploration of these maps. Constructing and generating such bibliometric networks graphically needs a consistent set of raw data, or metadata tto be more specific. VOSviewer focuses on the existing links between the metadata, as they exist in the database of analysis.

Bibliometric indicators are important in the analysis of scientific performance output and are needed to develop a quantitative (e.g.: number of publications) and quantitative & qualitative (e.g.: number of publications and their corresponding citations) inventories of the publishing activity at many levels: at country-level, institution-level, author(s) (Pătărlăgeanu et al. 2020a). Not only can VOSviewer facilitate this type of analysis, but VOSviewer can also facilitate research efficiency in analyzing keywords in the title and abstract of the analyzed scientific publications.

Five queries were performed on the Web of Science database in April 2021. Thus, raw metadata extracted and processed in this quantitative research assessment

Table 2 The queries performed on the Web of Science database in April 2021

No	Query	Number of identified publications indexed in WoS
Query 1	TOPIC: ("sustainable consumption")	2,593
Query 2	TOPIC: ("sustainable production")	5,978
Query 3	TOPIC: ("sustainable consumption and production")	330
Query 4	TOPIC: ("sustainable consumption") AND TOPIC: ("sustainable production")	103
Query 5	TOPIC: ("SDG12")	93

Source Authors' development

includes only the scientific work published and indexed in the Web of Science at the moment the queries from Table 2 were performed on the Web of Science database. All five queries are specific to the topics of sustainable consumption, sustainable production and a mix of both. The following searching filter was automatically applied by the Web of Science platform in the case of all the performed queries: "Timespan: all years. Indexes: SCI-EXPANDED, SSCI, A & HCI, CPCI-S, CPCI-SSH, BKCI-S, BKCI-SSH, ESCI, CCR-EXPANDED, IC."

Based on the querying results from Table 2, the most suitable set of metadata for developing the assessment of the scientific publications elaborated on the topic of sustainable consumption and production is Query 3. In accordance with the objective of this research Web of Science is a valuable database, because it provides the premises for the development of an impactful research assessment in the fields of sustainable consumption and production, having all the publications elaborated on these topics as the foundation of the analysis.

5 Results and Discussions

5.1 Performance Assessment Based on Eurostat SDG 12 Indicators

5.1.1 Per-Indicator and Per-Country Statistical Analysis

a. Resource productivity and domestic material consumption.

Table 3 highlights the performance of each EU-27 member with respect to the first analyzed SDG 12 indicator, based on the results from 2015 (the year the 2030 Agenda was published) and 2019 (the most recent year for which data were available in the Eurostat database). The improvement was looked upon through the lens of the absolute difference between the result tracked in 2019 and the one tracked in 2015.

Table 3 The progress assessment with regard to resource productivity and domestic material consumption (euro per kilogram, chain linked volumes—2015) in the EU-27

Country	Year 2015	Year 2019	2019 reported to 2015 Improvement (Absolute)	Improvent Ranking	Year 2019 Ranking	Performance Discrepancies
Austria	2.77	3.62	0.85	2	3	1
Belgium	0.30	0.36	0.06	17	27	10
Bulgaria	1.01	1.13	0.12	11	20	9
Croatia	2.10	2.10	0.00	23	11	-12
Cyprus	2.39	2.64	0.25	6	8	2
Czechia	0.58	0.64	0.05	19	25	6
Denmark	2.74	2.75	0.02	21	7	-14
Estonia	1.33	1.75	0.43	4	13	9
Finland	2.63	2.80	0.17	8	6	-2
France	2.97	3.04	0.07	14	5	-9
Germany	1.08	1.15	0.06	15	19	4
Greece	3.43	3.56	0.13	10	4	-6
Hungary	1.49	1.31	-0.19	27	17	-10
Ireland	0.96	1.04	0.08	13	21	8
Italy	0.86	0.82	-0.04	24	22	-2
Latvia	3.85	4.10	0.26	5	2	-3
Lithuania	0.90	0.77	-0.13	25	23	-2
Luxembourg	1.63	2.13	0.50	3	10	7
Malta	3.69	4.96	1.27	1	1	0
Netherlands	2.16	2.22	0.06	16	9	-7
Poland	0.67	0.77	0.10	12	24	12
Portugal	1.12	1.16	0.05	20	18	-2
Romania	0.36	0.37	0.01	22	26	4
Slovakia	1.42	1.60	0.19	7	14	7
Slovenia	1.18	1.33	0.15	9	15	6
Spain	1.26	1.32	0.06	18	16	-2
Sweden	2.02	1.86	-0.16	26	12	-14

Source Authors' development, based on Eurostat data (2021)

Based on this difference, the improved ranking was calculated: the country with the greatest recorded difference was allocated the greatest rank (1) and so on. Moreover, performance discrepancy was calculated as the difference between the 2019 ranking and the improvement ranking. In this regard, high positive values signal that there is a gap between the European country of analysis and the European leaders as far as resource productivity and domestic material consumption are concerned. High negative values signal the opposite: the country of analysis was an EU-27 leader already, ahead of the rest of the EU-27 members, and did not register impressive improvement based on the 2019–2015 calculated difference.

Results show that Malta, Austria and Luxembourg registered the greatest performance regarding the improvement made in the sense of increasing resource productivity in the EU-27. In the case of Malta, this is even more impressive, considering the fact that Malta scored the first rank in this regard in 2015, which repetead in 2019. Not only that, but Malta also registered the greatest improved in the EU-27: 1.27 euro per kilogram (increase of 34.41%, 2019 reported to 2015). Hungary, Sweden and Lithuania registered poor performance with respect to resource productivity: in the case of these countries, we actually did not see progress, but on the contrary: regress—this is because their resource productivity decreased in 2019. However, Sweden and Hungary managed to rank on the 12nd and 17th places in the EU-27, based on their resource productivity registered in 2019: 1.86 and 1.31 euro per kilogram. On the other hand, based on the improvement ranking, Lithuania was ranked 25/27, and, the level of the year 2019, the country was at the bottom of the ranking as well: 23.

b. Circular material use rate.

The progress with regard to the circular material use rate in the EU-27 was presented in Table 4. The greatest achievements were observed in the case of Austria (5.6%, from 18.4% in 2015 to 24.0% in 2019), Czechia (3.3%, from 11.8% in 2015 to 15.1% in 2019), Finland (2.7%, from 7.5% in 2015 to 10.2% in 2019). Austria was already one of the EU-27 leaders in terms of circular material use rate before the year 2019, but the progress made during the period of analysis made Austria an example to follow in the EU-27, considering all its efforts to improve the circularity characteristic of the Austrian economy.

In the case of eight EU-27 members, regress can be observed: Italy, Romania, Denmark, Spain, Croatia, Belgium, Ireland and Poland. The average of the improvement (EU-27) was 1.04%, while the previously mentioned countries registered an average decrease of the circular material use rate of 0.59%, with maximum points reached in the case of Poland: 1.8%. In 2019, at the bottom of this ranking was Romania (1.5% circularity rate), followed by Denmark (1.6%) and Portugal (2.2%). These countries could collaborate with Austria to find good practice examples and try adapting and replicating them, striving to attain progress.

c. Generation of hazardous waste, excluding major mineral wastes.

In Table 5, the progress with regard to the generation of hazardous waste, excluding major mineral wastes (kilograms per capita) was presented in the case of the EU-27 members. Under the form of a decrese of the nominal hazardous waste generated, the greatest achievements were observed in the case of Luxembourg (233 kg per capita), Latvia (100 kg per capita) and Austria (38 kg per capita). Unfortunately, only seven EU-27 countries registered improvements: Belgium, Cyprus, the Netherlands and Croatia, besides the previously mentioned top three European leaders.

Overall, the EU-27 generated more hazardous waste per capita in 2018 than in 2016 (26 more kilograms per capita, on average). The greatest regress was observed in the case of Czechia: 841 more kilograms of hazardous waste in 2018, followed by Poland and Spain. In 2018, the ranking of the least-hazardous-waste-generators in the EU-27 was the following: Romania on the first place, followed by Belgium, Hungary, Germany and Ireland.

Table 4 The progress assessment with regard to the circular material use rate in the EU-27

Country	Year 2015	Year 2019	2019 reported to 2015 Improvement (Absolute)	Improvent Ranking	Year 2019 Ranking	Performance Discrepancies
Austria	18.40%	24.00%	5.60%	1	2	1
Belgium	3.10%	2.40%	-0.70%	25	24	-1
Bulgaria	6.90%	8.30%	1.40%	11	12	1
Croatia	8.40%	7.80%	-0.60%	24	13	-11
Cyprus	11.60%	12.20%	0.60%	14	6	-8
Czechia	11.80%	15.10%	3.30%	2	5	3
Denmark	1.90%	1.60%	-0.30%	22	26	4
Estonia	1.90%	4.20%	2.30%	6	21	15
Finland	7.50%	10.20%	2.70%	3	10	7
France	18.70%	20.10%	1.40%	10	3	-7
Germany	4.30%	4.90%	0.60%	15	19	4
Greece	17.30%	19.30%	2.00%	9	4	-5
Hungary	2.40%	2.90%	0.50%	17	23	6
Ireland	5.40%	4.70%	-0.70%	26	20	-6
Italy	4.10%	4.00%	-0.10%	20	22	2
Latvia	9.70%	11.90%	2.20%	7	7	0
Lithuania	5.80%	6.80%	1.00%	13	16	3
Luxembourg	4.60%	7.10%	2.50%	5	14	9
Malta	25.80%	28.50%	2.70%	4	1	-3
Netherlands	11.00%	11.50%	0.50%	16	8	-8
Poland	11.60%	9.80%	-1.80%	27	11	-16
Portugal	2.10%	2.20%	0.10%	19	25	6
Romania	1.70%	1.50%	-0.20%	21	27	6
Slovakia	8.40%	10.40%	2.00%	8	9	1
Slovenia	5.00%	6.10%	1.10%	12	18	6
Spain	6.50%	6.20%	-0.30%	23	17	-6
Sweden	6.80%	7.00%	0.20%	18	15	-3

Source Authors' development, based on Eurostat data (2021)

d. Generation of non-hazardous waste, excluding major mineral wastes.

The progress with regard to the reduction of the generation of non-hazardous waste in the EU-27 was presented in Table 6. The greatest achievements were observed in the case of Latvia (decrease of 319 kg per capita, from 2,328 in 2015 to 2,009 in 2019), Denmark (decrease of 174 in the same timeframe), and Czechia (decrease of 97 kg per capita).

In the case of seventeen EU-27 members, regress was observed, especially in the case of Belgium, Italy and Austria. The average regress at the level of EU-27 was 41 kg per capita, while the previously mentioned group of seventeen countries registered an average increase of nominal non-hazardous waste of 113 kg per capita. The most unfavorable situation was registered in the case of Belgium, where the nominal non-hazardous waste generated increased with 581 kg per capita (increase of 23.37%, 2018 reported to 2016). Poor performing countries should collaborate

Table 5 The progress assessment with regard to the generation of hazardous waste, excluding major mineral wastes (kilograms per capita)

Country	Year 2016	Year 2018	2018 reported to 2016 Improvement (Absolute)	Improvent Ranking	Year 2018 Ranking	Performance Discrepancies
Austria	245	207	-38	3	25	22
Belgium	41	30	-11	4	2	-2
Bulgaria	81	93	12	20	14	-6
Croatia	96	95	-1	7	15	8
Cyprus	176	174	-2	5	22	17
Czechia	7,345	8,186	841	27	27	0
Denmark	100	121	21	23	18	-5
Estonia	45	56	11	18	8	-10
Finland	64	65	1	8	11	3
France	111	123	12	21	19	-2
Germany	32	37	5	13	4	-9
Greece	144	152	8	16	21	5
Hungary	33	35	2	9	3	-6
Ireland	32	38	6	14	5	-9
Italy	45	49	4	11	6	-5
Latvia	369	269	-100	2	26	24
Lithuania	42	49	7	15	7	-8
Luxembourg	294	61	-233	1	10	9
Malta	133	144	11	19	20	1
Netherlands	115	113	-2	6	17	11
Poland	37	79	42	26	13	-13
Portugal	76	100	24	24	16	-8
Romania	20	22	2	10	1	-9
Slovakia	53	57	4	12	9	-3
Slovenia	65	74	9	17	12	-5
Spain	149	184	35	25	23	-2
Sweden	180	200	20	22	24	2

Source Authors' development, based on Eurostat data (2021)

with Latvia to find measures designed to limit the amount non-hazardous waste generated.

e. Gross value added in environmental goods and services sector.

In Table 7, the progress with regard to the gross value added (percentage of GDP) in environmental goods and services sector (kilograms per capita) was presented in the case of the EU-27 members. The greatest achievements were observed in the case of Luxembourg (233 kg per capita), Latvia (100 kg per capita) and Austria (38 kg per capita). Unfortunately, only seven EU-27 countries registered improvements: Luxembourg (0.75% increase) and Austria (0.68% increase), followed by Denmark, Ireland, Germany and Lithuania (average increase of 0.16% in the case of these four European countries).

Table 6 The progress assessment with regard to the generation of non-hazardous waste, excluding major mineral wastes (kilograms per capita)

Country	Year 2016	Year 2018	2018 reported to 2016 Improvement (Absolute)	Improvent Ranking	Year 2018 Ranking	Performance Discrepancies
Austria	3,138	3,297	159	25	27	2
Belgium	2,486	3,067	581	27	26	-1
Bulgaria	1,133	1,179	46	14	7	-7
Croatia	1,560	1,679	119	22	17	-5
Cyprus	1,721	1,698	-23	7	18	11
Czechia	1,621	1,524	-97	3	16	13
Denmark	1,664	1,490	-174	2	14	12
Estonia	1,281	1,422	141	23	11	-12
Finland	1,416	1,475	59	17	13	-4
France	1,337	1,381	44	13	10	-3
Germany	818	885	67	19	2	-17
Greece	1,652	1,698	46	15	19	4
Hungary	812	895	83	20	3	-17
Ireland	716	663	-53	5	1	-4
Italy	1,188	1,354	166	26	9	-17
Latvia	2,328	2,009	-319	1	22	21
Lithuania	1,077	1,051	-26	6	5	-1
Luxembourg	983	1,029	46	16	4	-12
Malta	2,406	2,469	63	18	25	7
Netherlands	1,772	1,770	-2	10	20	10
Poland	2,053	2,032	-21	8	23	15
Portugal	1,071	1,215	144	24	8	-16
Romania	1,064	1,087	23	12	6	-6
Slovakia	1,405	1,422	17	11	12	1
Slovenia	1,395	1,505	110	21	15	-6
Spain	2,447	2,375	-72	4	24	20
Sweden	1,956	1,935	-21	9	21	12

Source Authors' development, based on Eurostat data (2021)

As far as this sustainable consumption and production-specific indicator is concerned, regress was observed in the case of nine EU-27 members: Romania, Latvia, Czechia, Portugal, Croatia, Malta, Sweden, Spain and Italy (average decrease of 0.16% of the gross value added, as percentage of GDP, in environmental goods and services sector). However, at the level of EU-27, an average small increase was registered in 2018: 0.04%.

Table 7 The progress assessment with regard to the gross value added (percentage of GDP) in environmental goods and services sector

Country	Year 2015	Year 2018	Improvent Ranking	Improvent Ranking	Year 2018 Ranking	Performance Discrepancies
Austria	3.62%	4.30%	0.68%	2	3	1
Belgium	0.94%	0.94%	0.00%	14	22	8
Bulgaria	1.82%	1.90%	0.08%	9	16	7
Croatia	1.59%	1.45%	-0.14%	19	20	1
Cyprus	N/A	N/A	N/A	N/A	N/A	N/A
Czechia	2.49%	2.30%	-0.19%	21	8	-13
Denmark	2.99%	3.19%	0.20%	3	4	1
Estonia	4.40%	4.45%	0.05%	12	2	-10
Finland	5.63%	5.69%	0.06%	10	1	-9
France	1.54%	1.62%	0.08%	7	18	11
Germany	1.81%	1.96%	0.15%	5	15	10
Greece	N/A	N/A	N/A	N/A	N/A	N/A
Hungary	N/A	N/A	N/A	N/A	N/A	N/A
Ireland	0.72%	0.88%	0.16%	4	23	19
Italy	1.90%	1.87%	-0.03%	15	17	2
Latvia	2.78%	2.53%	-0.25%	22	6	-16
Lithuania	2.08%	2.20%	0.12%	6	13	7
Luxembourg	1.68%	2.43%	0.75%	1	7	6
Malta	1.13%	0.99%	-0.14%	18	21	3
Netherlands	2.17%	2.25%	0.08%	8	10	2
Poland	2.16%	2.21%	0.05%	13	12	-1
Portugal	2.44%	2.28%	-0.16%	20	9	-11
Romania	3.40%	3.00%	-0.40%	23	5	-18
Slovakia	N/A	N/A	N/A	N/A	N/A	N/A
Slovenia	1.55%	1.60%	0.05%	11	19	8
Spain	2.27%	2.22%	-0.05%	16	11	-5
Sweden	2.20%	2.08%	-0.12%	17	14	-3

Source Authors' development, based on Eurostat data (2021)

5.2 *Overall Perspective on the Progress and Performance in the EU-27 with Regard to Selected SDG 12 Indicators*

Pivoting the previously discussed indicators into a single table with the ranking perspective in the spotlight of the research was done in Table 8. Results highlight that Austria, Malta and Latvia follow the same performant sustainable consumption and production pattern as far as the circular material use rate is concerned, as well as resource productivity. However, Austria is facing issues in terms of the amount of waste generated per capita, which is opening the scene for other European countries for performance, such as Ireland, Germany and Hungary. However, looking at the gross value added (percentage of GDP) in environmental goods and services sector,

Table 8 The multinational (EU-27) perspective on the progress and performance in the EU-27 with regard to selected SDG 12 indicators

Country	Resource productivity and domestic material consumption		Circular material use rate		Generation of hazardous waste		Generation of non-hazardous waste		Gross value added in environmental goods and services sector	
	IR*	RML**	IR*	RML**	IR*	RML**	IR*	RML**	IR*	RML**
Austria	2	3	1	2	3	25	25	27	2	3
Belgium	17	27	25	24	4	2	27	26	14	22
Bulgaria	11	20	11	12	20	14	14	7	9	16
Croatia	23	11	24	13	7	15	22	17	19	20
Cyprus	6	8	14	6	5	22	7	18	N/A	N/A
Czechia	19	25	2	5	27	27	3	16	21	8
Denmark	21	7	22	26	23	18	2	14	3	4
Estonia	4	13	6	21	18	8	23	11	12	2
Finland	8	6	3	10	8	11	17	13	10	1
France	14	5	10	3	21	19	13	10	7	18
Germany	15	19	15	19	13	4	19	2	5	15
Greece	10	4	9	4	16	21	15	19	N/A	N/A
Hungary	27	17	17	23	9	3	20	3	N/A	N/A
Ireland	13	21	26	20	14	5	5	1	4	23
Italy	24	22	20	22	11	6	26	9	15	17
Latvia	5	2	7	7	2	26	1	22	22	6
Lithuania	25	23	13	16	15	7	6	5	6	13
Luxembourg	3	10	5	14	1	10	16	4	1	7
Malta	1	1	4	1	19	20	18	25	18	21
Netherlands	16	9	16	8	6	17	10	20	8	10
Poland	12	24	27	11	26	13	8	23	13	12
Portugal	20	18	19	25	24	16	24	8	20	9
Romania	22	26	21	27	10	1	12	6	23	5
Slovakia	7	14	8	9	12	9	11	12	N/A	N/A
Slovenia	9	15	12	18	17	12	21	15	11	19
Spain	18	16	23	17	25	23	4	24	16	11
Sweden	26	12	18	15	22	24	9	21	17	14

*Improvement Ranking; **Ranking at Multinational (EU-27) Level (2018 or 2019, based on data availability)
Source Authors' development, based on Eurostat data (2021)

Austria is once again one of the European leaders in terms of performance, as well as Estonia and Finland.

Looking at the results presented in Table 8, one can notice that the EU-27 countries that: (a) registered the greatest progress with regard to SDG 12 and (b) got placed in the top of the performance ranking at multinational level are the following: (1) Luxembourg; (2) Finland; (3) Austria; (4) Latvia; (5) Slovakia. From this quantitative research approach, at the bottom of the ranking are Belgium, Portugal, Sweden, Spain and Italy.

The assessment on EU-27's progress and performance with regard to SDG 12 was also approached through the lens of the scientific work carried out and promoted in prestigious journals and proceedings by resorting to the bibliometric analysis (Fig. 1).

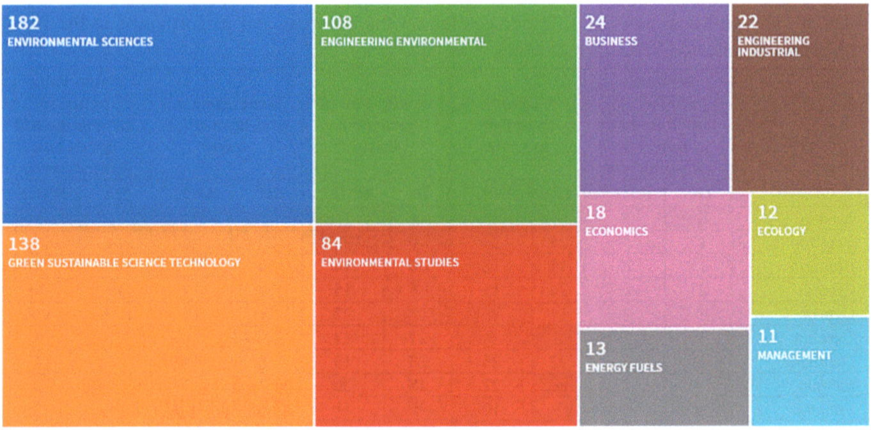

Fig. 1 Treemap of the top 10 WoS categories associated to the 330 identified publications. *Source* web of science

5.3 *Bibliometric Analysis Based on Raw Web of Science Results*

According to the results based on the performed query, bibliometric findings referring to the Web of Science categories associated to the 330 identified publications show that *Environmental Sciences* represents the most commonly associated category (182 publications; 55.15%), followed by *Green Sustainable Science Technology* (138 publications; 41.81%) and *Engineering Environmental* (108 publications, 32.72%). The 330 indexed publications had 750 Web of Science categories associated, which is a usual situation: one publication can have more than one category associated (Leydesdorff et al. 2013).

Figures 2 and 3 explain the intensification of the scientific interest after the year 2014 for the topics of sustainable consumption and production patterns, especially considering that the 2030 Agenda for Sustainable Development was published in 2015. Consequently, the immediate scientific response to the development paradigm shift is observed through the numerous papers elaborated on the development policy, a modern one, which has the sustainability and wealthiest factors at its core.

Regarding the assessment on EU-27's progress and performance with regard to SDG 12, approached through the lens of the scientific work carried out, published and indexed in the Web of Science, it is important to mention the Kingdom of the Netherlands as the leader in the EU-27 as far as number of publications is concerned, as well as the number of citations. Based on the data included in Table 9, Germany, France, Sweden and Italy are also in the top of the ranking assessing the scientific progress with regard to research towards SDG 12.

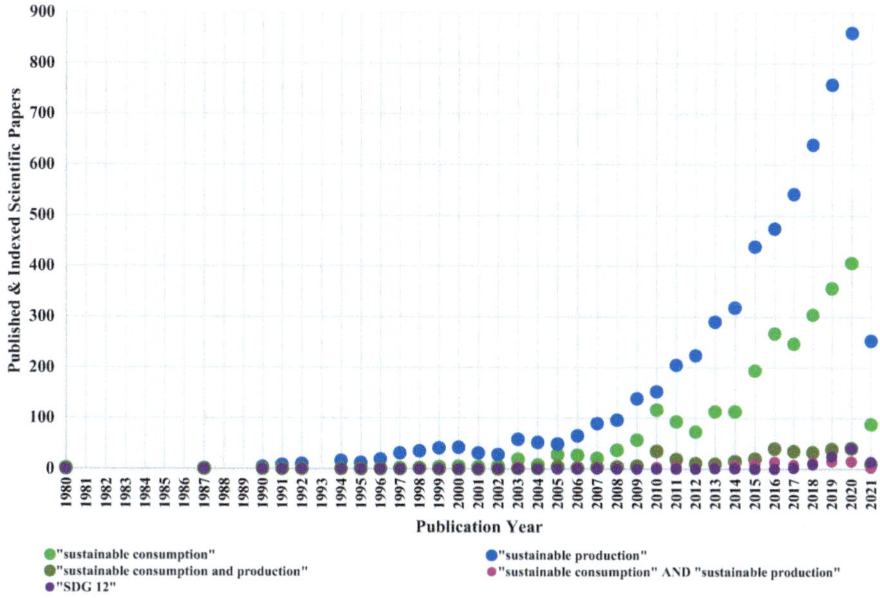

Fig. 2 The evolution of the yearly published and indexed scientific papers approaching the topics of sustainable consumption and production patterns. *Source* web of science

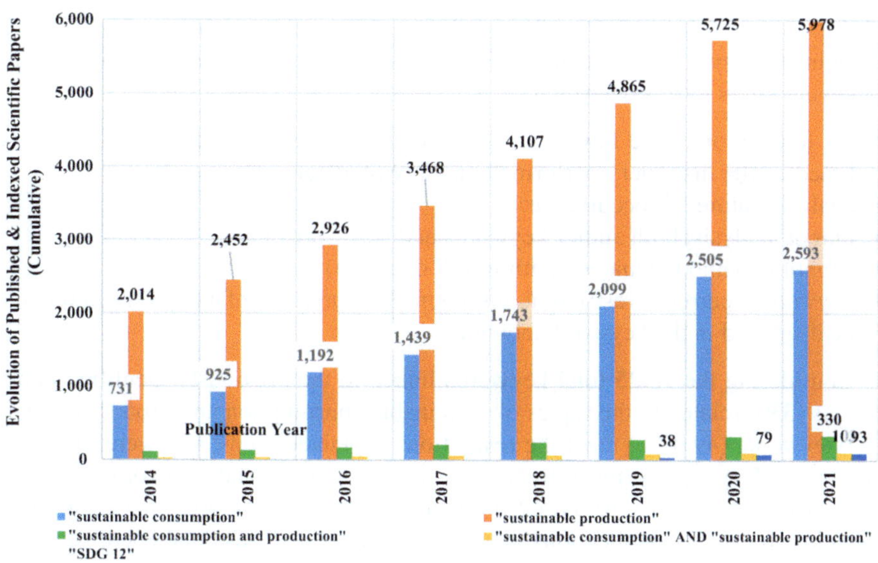

Fig. 3 Cumulative perspective on the evolution of the published and indexed scientific papers approaching the topics of sustainable consumption and production patterns. *Source* web of science

Table 9 The ranking of the top 10 countries having authors publishing scientific results on the topics of "sustainable consumption and production" (Query 3, as explained in Table 2)

Top 10 Countries	Number of publications	Share from the total publication (%)	Ranking based on the number of publications	Citations	Share from the total citations (%)	Ranking based on the number of citations
England	51	17.41	1	2,372	23.06	1
Netherlands	41	13.99	2	2,032	19.75	2
China	37	12.63	3	1,053	10.24	4
USA	35	11.95	4	978	9.51	5
Germany	32	10.92	5	1,284	12.48	3
France	22	7.51	6	676	6.57	7
Japan	22	7.51	7	191	1.86	10
Sweden	21	7.17	8	738	7.17	6
Italy	17	5.80	9	448	4.35	9
Denmark	15	5.12	10	516	5.02	8

Source Web of science

5.4 *Bibliometric Analysis Based on Web of Science Results Processed in VOSviewer*

Mapping the keyword correlation in Fig. 4 was made possible based on applying the following filters in VOSviewered:

- the foundation of the raw data is represented by the 237 publications identified based on the first query performed on the WoS database;
- analysis nature: co-occurrence;
- unit of analysis: both author keywords and KeyWords Plus;
- counting method: equal weighting (full counting);
- keyword occurrence restriction: minimum 3 occurrences of a keyword in order to be displayed in the keyword network.

Considering the 1,525 total keywords linked to the previously identified publications (330) related to the topics of sustainable consumption and production, only 162 keywords (49.09%) meet the restrictions explained above and were displayed in Fig. 4. The top 20 most frequently used keywords and keyword structures were analyzed in Table 10, based on their occurrence in the raw metadata corresponding to the 237 publications identified based on the "sustainable consumption and production" Web of Science query.

The evolution during 2014–2020 show that between 2014 and 2016, the keywords specific to the publications dealing with sustainable consumption and production were: "policy instruments", "models", "selection", "growth", "consumer behavior", "emissions", "product service systems", "science", "ecological footprint", "household consumption", "determinant", "fair trade", "networks", "policy", "recovery",

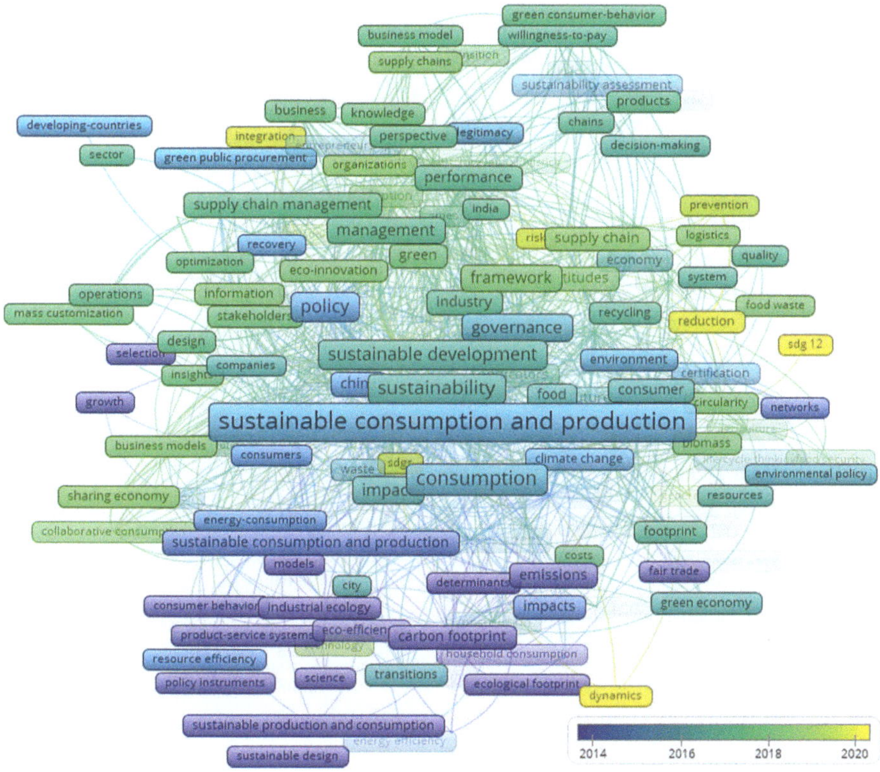

Fig. 4 The evolution of the usage of the keywords linked to the 330 publications identified based on the "sustainable consumption and production" Web of Science query. *Source* Authors' development in VOSviewer 1.16.16

"developing countries", "green public procurement"; while between 2017 and 2020 the dynamic of the matter changed from treating sustainable consumption and production as strategy into building a legislative framework dedicated to consolidate the transition from the conventional consumption and production models, as suggested by the following identified keywords: "eco-innovation", "mass customization", "optimization", "knowledge", "perspective", "business model", "willingness to-pay", "products", "chains", "green consumer behavior", "management", "governance", "framework", "sharing economy", "collaborative consumption", "business models", "integration", "reduction", "dynamics", "sdg 12", "risk", "logistics", "green" and "supply chains".

In the scientific papers analyzed based on the "sustainable consumption and production" Web of Science query, the literature references a new method to align the world in an innovative model that can sustain a better life quality in a more sustainable and less harming manner for the environment. This model inolves circularity and is based on the 'green'-oriented economy, involving strong partnerships (Fig. 5).

Table 10 Ranking the frequency of keyword occurrence

Ranking	Keyword/keyword structure	Occurrences	The share of occurrences from the total of top 20 (%)
1	Sustainable consumption and production	81	17.23
2	Consumption	37	7.87
3	Sustainability	34	7.23
4	Sustainable consumption	31	6.60
5	Sustainable development	27	5.74
6	Policy	26	5.53
7	Governance	23	4.89
8	Impact	21	4.47
9	Circular economy	20	4.26
10	Management	19	4.04
11	Systems	19	4.04
12	Energy	18	3.83
13	Framework	16	3.40
14	Sustainable development goals	15	3.19
15	Behavior	14	2.98
16	China	14	2.98
17	Industry	14	2.98
18	Performance	14	2.98
19	Sustainable consumption and production (scp)	14	2.98
20	Life cycle assessment	13	2.77

Source Authors' development

The red keyword cluster from Fig. 5: "operations", information "growth", "consumer behavior", "sharing economy", "optimization", "models", "product-service systems", "carbon footprint", "ecological footprint", "resource" "efficiency" "policy instrument", "eco-efficient" closely illustrates the direct link between the impact of a reduced ecological footprint business models and the potential for economic growth, while taking into consideration the opportunity created by sustainable models to reduce the intensity of the usage of natural resources, calling for their proper management.

Fig. 5 Cluster analysis of the keywords linked to the 330 publications identified based on the "sustainable consumption and production" Web of Science query. *Source* Authors' development in VOSviewer 1.16.16

The orange cluster: "business models", "supply chains", "perspective", "performance", "chains", "decision-makers", "sustainability assessment", "products" "willingness to-pay", "green consumer behavior" highlights the need for transitioning from the conventional consumption and production models to a more sustainable alternative by creating green consumer behavior and resilient supply chains.

The green cluster: "mass customization", "design", "stakeholders", "companies", "business models", "consumers", "sustainable development", "chain", "sdgs", "sustainable consumption and production", "city, "sustainable design", "sustainability, "dynamics", "costs", "fair trade", "green economy", "climate change", "sustainability", "industry, "logistics" is focused on the principle of fairness in relation with the profitability of the small business owns. This cluster signals the importance of consolidating industries more sustainability, in order to create a equitable equilibrium between costs and prices, in the light of the established goals of the 2030 Agenda for Sustainable Development.

The blue cluster: "framework"; "risk"; "supply chain", "consumption", "impacts", "industrial ecology", "emissions", "household consumption", "resources", "circularity", "networks", "transitions", "science", "technology", "food waste", "supply chain", "economy", "recycling", "prevention", "reduction", "sdg 12" illustrates the necessity of cooperation in different sectors, resorting to various industry and technologies, in order to ensure and promote the transition towards a more circular and inclusive economy.

Not only did the bibliometric analysis help assessing the progress and performance with regard to SDG 12 through the perspectives of the quantitative analysis of the existing publications and their citations, but the bibliometric analysis also helped mapping the most essential aspects of sustainable consumption and production patterns in the literature.

6 Conclusion

Implementing the 2030 Agenda for Sustainable Development successfully and efficiently is an imperative that calls for mutual understanding and collective global effort. Addressing the sustainability challenge simultaneously with achieving economic prosperity and eradicating poverty globally can be made possible if respecting and living in accordance with the principles instilled in the Agenda, and, probably more important, act and behave sustainably.

Although the research results might seem optimistic, the public health and socio-economic crisis caused by the COVID-19 pandemic serves as a negative influence on meeting most of the SDGs in a timely manner. However, with respect to the SDG 12, the pandemic's impact on its specific targets is still unclear. On the one hand, if approaching the positive outcome, the reduced intensity of economic activities and the short-term diminished consumption rate at the beginning of the pandemic in 2020—both of them caused an initial reduction in the use of natural resource, therefore relieving some global market tension. On the other hand, if approaching the negative outcome, resorting to plastic as a material for the production of personal protective equipment represents one of the production pandemic-specific behavior that has caused an increase in plastic pollution all over the world. In this context, the crisis represents an actual setback for SDG 12 ambitions, and, consequently, multiple research questions arose: has the decline in economic activities brought more environmental benefits at the beginning of the pandemic than the damage caused by the huge production of plastic-based protection sanitary equipment? Although correctly anticipating the long-term COVID-19 pandemic consequences is not likely due to high levels of uncertainty, another emerging research question is whether the nature of economic activities will return to the old patterns specific to exacerbated economic competitiveness and environmental degradation; will the pandemic production patterns persist in the post-pandemic world, or will global markets impose an increase of the production intensity when the pandemic is over? All these questions can represent the starting point of future research in the area of sustainable consumption and production patterns. Moreover, answering the previously mentioned research questions would represent the assessment of the resilience performance with regard to meeting the targets and indicators specific to SDG 12.

The outcome of this empirical analysis consolidates forthcoming works approaching the performance made towards SDG 12 by providing a methodological framework that is easily replicable once data referring to the year 2020 become available. Moreover, this research can be extended and more indicators can be included in the proposed framework.

References

Allen C, Metternicht G, Wiedmann T (2018) Initial progress in implementing the sustainable development goals (SDGs): a review of evidence from countries. Sustain Sci 13(5):1453–1467. https://doi.org/10.1007/s11625-018-0572-3

Aragaw TA (2020) Surgical face masks as a potential source for microplastic pollution in the COVID-19 scenario. Mar Pollut Bull 159:111517. https://doi.org/10.1016/j.marpolbul.2020.111517

Bocken NMP, Short SW, Rana P, Evans S (2014) A literature and practice review to develop sustainable business model archetypes. J Clean Prod 65:42–56. https://doi.org/10.1016/j.jclepro.2013.11.039

Bracale R, Vaccaro CM (2020) Changes in food choice following restrictive measures due to Covid-19. Nutr Metab Cardiovasc Dis 30(9):1423–1426. https://doi.org/10.1016/j.numecd.2020.05.027

Constantin M, Radulescu ID, Andrei JV, Chivu L, Erokhin V, Gao T (2021) A perspective on agricultural labor productivity and greenhouse gas emissions in the context of the common agricultural policy exigencies. Econ Agric 68(1):53–67. https://doi.org/10.5937/ekoPolj2101053C

Diaz-Sarachaga JM, Jato-Espino D, Castro-Fresno D (2018) Is the sustainable development goals (SDG) index an adequate framework to measure the progress of the 2030 Agenda? Sustain Dev 26(6):663–671. https://doi.org/10.1002/sd.1735

Dinu M, Pătărlăgeanu SR, Chiripuci B, Constantin M (2020a) Accessing the European funds for agriculture and rural development in Romania for the 2014–2020 period. Proc Int Conf Bus Excell 14(1):717–727. https://doi.org/10.2478/picbe-2020-0068

Dinu M, Pătărlăgeanu SR, Petrariu R, Constantin M, Potcovaru A-M (2020b) Empowering sustainable consumer behavior in the EU by consolidating the roles of waste recycling and energy productivity. Sustainability 12(23):9794. https://doi.org/10.3390/su12239794

Drăgoi MC, Popescu M-F, Andrei JV, Mieilă M (2018) Developments of the circular economy in romania under the new sustainabilty paradigm. Econ Comput Econ Cybernet Stud Res 52(2)

Eurostat (2021) Eurostat databases online

Evans S, Gregory M, Ryan C, Bergendahl MN, Tan A (2009) Towards a sustainable industrial system: with recommendations for education, research, industry and policy. University of Cambridge, Institute for Manufacturing

Fader M, Cranmer C, Lawford R, Engel-Cox J (2018) Toward an understanding of synergies and trade-offs between water, energy, and food SDG targets. Front Environ Sci 6 https://doi.org/10.3389/fenvs.2018.00112

Gasper D, Shah A, Tankha S (2019) The framing of sustainable consumption and production in SDG 12. Global Pol 10(S1):83–95. https://doi.org/10.1111/1758-5899.12592

Gray RS (2020) Agriculture, transportation, and the COVID-19 crisis. Can J Agric Econ/Rev Can D'agroeconomie 68(2):239–243. https://doi.org/10.1111/cjag.12235

Hamari J, Sjöklint M, Ukkonen A (2016) The sharing economy: why people participate in collaborative consumption. J Am Soc Inf Sci 67(9):2047–2059. https://doi.org/10.1002/asi.23552
Hood WW, Wilson CS (2001) The literature of bibliometrics, scientometrics, and informetrics. Scientometrics 52(2):291. https://doi.org/10.1023/A:1017919924342
Ignat R, Constantin M (2020a) Multidimensional facets of entrepreneurial resilience during the COVID-19 crisis through the lens of the wealthiest Romanian counties. Sustainability 12(23):10220. https://doi.org/10.3390/su122310220
Ignat R, Constantin M (2020b) Short-term effects of COVID-19 pandemic on agri-food value chains in Romania. Innov Models Rev Glob Econ Sciendo 578–588. https://doi.org/10.2478/9788395815072-058
Ignat R, Constantin M (2021) Biowaste recycling as a solution towards a greener economy in the European Union. In: Proceedings of the international conference 'sustainable agriculture and rural development'. Belgrade, Serbia, pp 47–56
Istudor N, Ion RA, Petrescu IE, Hrebenciuc A (2019) Agriculture and the twofold relationship between food security and climate change evidence from Romania. Amfiteatrul Econ 21(51):285–285
Jackson T, Senker P (2011) Prosperity without growth: economics for a finite planet. Energy Environ 22(7):1013–1016. https://doi.org/10.1260/0958-305X.22.7.1013
Lebel L, Lorek S (2008) Enabling sustainable production-consumption systems. Annu Rev Environ Resour 33(1):241–275. https://doi.org/10.1146/annurev.environ.33.022007.145734
Leydesdorff L, Carley S, Rafols I (2013) Global maps of science based on the new Web-of-Science categories. Scientometrics 94(2):589–593. https://doi.org/10.1007/s11192-012-0784-8
Mekonnen MM, Hoekstra AY (2012) A global assessment of the water footprint of farm animal products. Ecosystems 15(3):401–415. https://doi.org/10.1007/s10021-011-9517-8
Mont O, Plepys A (2008) Sustainable consumption progress: should we be proud or alarmed? J Clean Prod 16(4):531–537. https://doi.org/10.1016/j.jclepro.2007.01.009
Narayanaswamy V, Stone L (2007) From cleaner production to sustainable production and consumption in Australia and New Zealand: achievements, challenges, and opportunities. J Clean Prod 15(8):711–715. https://doi.org/10.1016/j.jclepro.2006.06.011
Papargyropoulou E, Lozano R, Steinberger K, Wright J, Ujang N, Bin Z (2014) The food waste hierarchy as a framework for the management of food surplus and food waste. J Clean Prod 76:106–115. https://doi.org/10.1016/j.jclepro.2014.04.020
Pătărlăgeanu SR, Dinu M, Constantin M (2020a) Bibliometric analysis of the field of green public procurement. Amfiteatru Economic 22(53):71–81
Pătărlăgeanu SR, Rădulescu CV, Dinu M, Constantin M, (2020b) The impact of heavy work investment on the economy and the individual. Amfiteatru Econ 22(SI 14):1085–1102
Peattie K (2010) Green consumption: behavior and norms. Annu Rev Environ Resour 35(1):195–228. https://doi.org/10.1146/annurev-environ-032609-094328
Peattie K, Peattie S (2009) Social marketing: a pathway to consumption reduction? J Bus Res 62(2):260–268. https://doi.org/10.1016/j.jbusres.2008.01.033
Pradhan P, Costa L, Rybski D, Lucht W, Kropp JP (2017) A systematic study of sustainable development goal (SDG) interactions. Earth's Fut 5(11):1169–1179. https://doi.org/10.1002/2017EF000632
Røpke I (2009) Theories of practice—new inspiration for ecological economic studies on consumption. Ecol Econ 68(10):2490–2497. https://doi.org/10.1016/j.ecolecon.2009.05.015
Rowan NJ, Laffey JG (2020) Challenges and solutions for addressing critical shortage of supply chain for personal and protective equipment (PPE) arising from Coronavirus disease (COVID19) pandemic–case study from the Republic of Ireland. Sci Total Environ 725:138532. https://doi.org/10.1016/j.scitotenv.2020.138532
Roy V, Singh S (2017) Mapping the business focus in sustainable production and consumption literature: review and research framework. J Clean Prod 150:224–236. https://doi.org/10.1016/j.jclepro.2017.03.040

Sarkodie SA, Owusu PA (2020) Impact of COVID-19 pandemic on waste management. Environ Dev Sustain. https://doi.org/10.1007/s10668-020-00956-y

Selam MN (2020) Hand sanitizers marketed in the streets of addis ababa, Ethiopia, in the Era of COVID-19: a quality concern. Risk Manag Healthcare Pol 13:2483–2487. https://doi.org/10.2147/RMHP.S284007

Terama E, Milligan B, Jiménez-Aybar R, Mace GM, Ekins P (2016) Accounting for the environment as an economic asset: global progress and realizing the 2030 Agenda for sustainable development. Sustain Sci 11(6):945–950. https://doi.org/10.1007/s11625-015-0350-4

Thøgersen J, Ölander F (2002) Human values and the emergence of a sustainable consumption pattern: a panel study. J Econ Psychol 23(5):605–630

United Nations (2015) Transforming our world: the 2030 Agenda for sustainable development A/RES/70/1

van Eck NJ, Waltman L (2010) Software survey: VOSviewer, a computer program for bibliometric mapping. Scientometrics 84(2):523–538. https://doi.org/10.1007/s11192-009-0146-3

van Eck NJ, Waltman L (2019) VOSviewer manual. Univeristeit Leiden, Leiden, pp 1–53

van Noordwijk M, Duguma LA, Dewi S, Leimona B, Catacutan DC, Lusiana B, Öborn I, Hairiah K, Minang PA (2018) SDG synergy between agriculture and forestry in the food, energy, water and income nexus: reinventing agroforestry? Curr Opin Environ Sustain 34:33–42. https://doi.org/10.1016/j.cosust.2018.09.003

Voicu-Dorobanțu R, Volintiru C, Popescu M-F, Nerău V, Ștefan G (2021) Tackling complexity of the just transition in the EU: evidence from Romania. Energies 14(5):1509. https://doi.org/10.3390/en14051509

Volintiru CA, Popescu M-F, Frantescu D, Ciot M-G (2019) Political support at EU level for energy and environmental policies. Roman J Eur Aff 19(2):30–50

Sustainable Production-Consumption of the Microgrids: An Optimal Approach for Industrial Area

Hêriş Golpîra, Salah Bahramara, and Erfan Babaee Tirkolaee

Abstract Supplying the electrical energy demand of the industrial areas through the microgrid (MG) structure can increase the sustainable development indices in the electrical energy systems. Some of the MG's load is met through renewable energy sources (RESs) regarding which the pollution emission of the system decreases. Also, the MG operator (MGO) can decrease the operation cost of the system using the optimal scheduling of the electrical energy storages (EESs) and the shiftable loads. In the end, since there are several energy resources in the MG, the MGO can meet the MG's demand with a minimum amount of the energy not supplied (ENS). In this paper, a tri-objective function optimization model is developed for the energy management problem of an MG to minimize the operation cost, the pollution emission, and the ENS. The weighted goal programming approach is used to transform the proposed tri-objective function into a mixed-integer linear programming (MILP) model which can be solved through the CPLEX solver. To investigate the effectiveness of the proposed model and its solution approach, it is applied to an industrial MG. The results show that the MGO makes optimal decisions to schedule the MG's resources.

Keywords Sustainable energy management · Microgrids · Renewable energy sources · Tri-objective functio · Weighted goal programming

1 Introduction

Burning fossil fuel energy resources to produce electrical energy has a large share in the green house gas (GHG) emissions over the world. The problem is growing regarding the increasing electrical demand, especially in the industry section (Golpîra et al. 2021a). As reported by the International Energy Agency (IEA), the industry

H. Golpîra (✉) · S. Bahramara
Department of Electrical Engineering, Sanandaj Branch, Islamic Azad University, Sanandaj, Iran

E. B. Tirkolaee
Department of Industrial and Systems Engineering, Istinye University, Istanbul 34010, Turkey

C. J. Chiappetta Jabbour and S. A. R. Khan (eds.), *Sustainable Production and Consumption Systems*, Industrial Ecology,
https://doi.org/10.1007/978-981-16-4760-4_2

section consumed 9362 TWh electrical energy in 2018, i.e., 42 percent of the total electrical energy consumption in the world (Electricity final consumption by sector 2018). On the other hand, supplying the electrical energy demand of the consumers through the centralized power plants decreases the reliability of the electrical energy systems (Golpîra et al. 2021b) especially in the industrial areas. In fact, occurring faults on the industrial feeders in the distribution networks may lead to electrical energy outage for the industrial loads. To manage these challenges, two main solutions are proposed. The first one is producing electrical energy through renewable energy sources (RESs), recently studied in (Golpîra et al. 2018; Golpîra 2020) and the second one is supplying the local demand especially in industrial areas through local energy resources such as the distributed generations (DGs) and electrical energy storages (EESs). To implement these solutions, the structure of the electrical distribution networks changed with the introduction of microgrids (MGs) (Bahramara et al. 2020). The MGs are small electrical energy systems with local energy resources and demand that geographically limited to a certain domain in distribution networks. The MG Operator (MGO) can supply the MG's demand with the aim of achieving the sustainable indices through optimal scheduling of energy resources and shiftable loads (SLs). For this purpose, the MGO decides on the scheduling of DGs, EESs, and SLs to minimize the operation cost, the emission pollution, and the energy not supplied (ENS). Therefore, the aim of this paper is to develop an optimization model for the MGO to sustainable scheduling of its resources. This model can be applied in the industrial areas to supply their electrical energy demand.

2 Literature Review and Contributions

The energy management (EM) problem of an industrial MG is investigated by Daneshvar et al. (2021) using the Info-Gap Decision Theory (IGDT). The MG is equipped with the EES and RESs, and the risk-aversion level of the MGO to supply the demand taking into account the uncertainties is modeled in that research. A multi-objective optimization problem is developed by Choobineh and Mohagheghi (2016) to model the operation problem of an industrial MG where the MGO aims to minimize the operation cost, the pollution emission, and increasing the lifetime of MG's asset. The model predictive control approach is proposed by Bernardi et al. (2021) to model the EM problem of an industrial MG when a fault occurs in the network. Restrepo et al. (Restrepo et al. 2021) compared two energy management systems (EMSs), i.e., rule-based and optimization-based ones, for the MG real test system in Canada. The results revealed that the optimization-based EMS has the better performance for this system. Different deep learning algorithms are compared by Nakabi and Toivanen (2021) for the EMS of an MG includes a wind turbine, the EESs, and price-responsive loads. The flexibility of an industrial MG increases through the thermal energy storage and demand-side management strategies in (Carducci et al. 2017). Different algorithms are used to solve the EMS problem of three MG test systems in (Dey et al. 2019). The MGs include the RESs, the EES, and the micro-turbine and the MGO aims to minimize the operation cost of the system considering the uncertainties. A modified bat algorithm is used by Luo et al. (2020) to model the EMS problem of the MGs considering the uncertainties of the RESs and demand. The operation problem of an MG including the RESs, DGs, and EESs is modeled using different cost indices in (Adefarati et al. 2021). The results of this research showed the effectiveness of the proposed MG for decreasing the operation cost and increasing the sustainability of energy. The optimal dispatching of MG's energy resources in industrial zones is done using the stochastic optimization approach in (Zhang et al. 2018). The Markov decision process approach is used in this study to deal with the uncertainties of the RESs and electric vehicles. To obtain the economic and environmental goals in the MG's operation problem, the demand response programs and plug-in electric vehicles are employed in the proposed model in (Guo et al. 2021). In this study, a two-stage stochastic optimization approach is developed to model the uncertainties of the RESs, demand, energy price, and electric vehicles. Planning the MG consisting of the photo-voltaic arrays, the EES, and DGs is done by Lagrange et al. (2020) to decrease the total cost of the system and improve the resilience energy. The optimal planning problem of an MG is formulated to achieve the net-zero one in (Azimian et al. 2020). In this model, the demand shifting strategy is used to decrease the demand in peak hours. The MG's energy resources are optimally sizing in (Naderi et al. 2017) to obtain the sustainable and secure operation of the sensitive loads in an industrial area. The objective functions of this model are the total cost of the system, penetration rate of the RESs, and pollution emission. The operation problem of an MG is modeled as a robust optimization problem in (Bahramara and Golpîra 2018) where the uncertainty of the electric vehicles is managed in the decision-making problem of the MGO using the risk-aversion parameter as an important parameter in the process of decision-making under uncertainty (Golpîra 2018). A risk-based bi-level optimization approach is used in (Golpîra et al. 2020) to model the operation problem of an MG considering uncertainty. In this model, the MGO's risk-based decisions are formulated considering the conditional value at risk index. Although the economic and environmental indices are modeled in the proposed objective functions in the previous studies, the ENS index, which is an important index for the industrial areas, is not modeled besides the other objectives. Therefore, modeling the EM problem of the MGs considering all aspects of sustainability, i.e., economic, environmental, and social aspects, is not addressed in the previous studies. For this purpose, a tri-objective optimization model is developed in this paper for modeling the EM problem of the MGO considering sustainable indices. The objective functions aim to minimize the operation cost, the emission pollution, and the ENS.

3 Paper Organization

The rest of the paper is organized as follows. At first, the mathematical modeling of the proposed problem is presented and then its solution approach is described. Then, the numerical results are presented and in the last section, conclusions and future works are given.

4 Mathematical Modeling

The decision-making problem of the MGO is formulated as a multi-objective function in this section. The operation cost of the MG is minimized as (1). The first term models the total cost of the power trading of the MG with the grid and the second term models the operation cost of DGs.

$$\text{Minimize } OF^{\text{cost}} = \sum_t \rho_t (P_t^{\text{MG,in}} - P_t^{\text{MG,out}}) + \sum_t \sum_{k=1}^{K} C_k^{DG} P_{k,t}^{DG} \tag{1}$$

where ρ_t is the price of trading power with the market, $P_t^{\text{MG,in}}$ is the purchased power from the market, $P_t^{\text{MG,out}}$ is the sold power to the market, C_k^{DG} is the generation cost of DG, and $P_{k,t}^{DG}$ is the output power of DG k. Moreover, t and k indices are used for time and DGs, respectively.

The pollution emission of the MG is modeled as (2). The first term is related to the pollution emission produced through the main grid where if the MG sells energy to the grid, this pollution decreases. The pollution emission of the DGs is modeled as the second term of (2).

$$\text{Minimize } OF^{\text{Emission}} = \sum_t \gamma^{Grid} (P_t^{\text{MG,in}} - P_t^{\text{MG,out}}) + \sum_t \sum_{k=1}^{K} \gamma_k^{DG} P_{k,t}^{DG} \tag{2}$$

where γ^{Grid} is the emission factor of power generation of the main grid and γ_k^{DG} is the pollution factor of DGs.

The aim of the third objective function is to minimize the ENS of the system as modeled in (3).

$$\text{Minimize } OF^{\text{Social}} = \sum_t E_t^{ENS} \tag{3}$$

where E_t^{ENS} is the energy not supplied.

The proposed tri-objective functions are optimized considering the following constraints:

- Power balance constraint: The power generation of DGs and PV, the power purchased from the grid, the discharging power of the EES, the amount of load that is shifted down, and the amount of the ENS that is considered as the production side (the right-hand side) of power balance constraint modeled in (4). The sold power to the grid, the amount of shifted-up demand, the power charging of the EES, and the MG's load are modeled in the consumption side (the left-hand side) of (4).

$$\sum_{k=1}^{K} P_{k,t}^{\mathrm{DG}} + P_t^{\mathrm{PV}} + P_t^{\mathrm{MG,in}} + P_t^{\mathrm{SL\_DN}} + P_t^{dis}\eta^{dis} + E_t^{\mathrm{ENS}} = P_t^{\mathrm{MG,out}} + P_t^{\mathrm{SL\_UP}} + \frac{P_t^{ch}}{\eta^{ch}} + P_t^L \quad \forall t \tag{4}$$

where P_t^{PV} is the power generation of photovoltaic arrays, $P_t^{\mathrm{SL\_DN}}$ is the amount of shifted-down of the demand,$P_t^{\mathrm{SL\_UP}}$ is the amount of shifted-up of the demand, P_t^{dis} is power discharging of the EES, P_t^{ch} is power charging of the EES, P_t^L is the demand of the MG, η^{dis} and η^{ch} are the efficiency of power discharging and charging of the EES, respectively.

- Constraints of MG's power trading with the main grid: Eqs. (5) and (6) are used to model the power trading power limitation of the MG with the main grid.

$$0 \le P_t^{MG,in} \le P^{MG,max} X_t^{MG} \quad \forall t \tag{5}$$

$$0 \le P_t^{MG,out} \le P^{MG,max}(1 - X_t^{MG}) \quad \forall t \tag{6}$$

where $P^{MG,max}$ is the maximum power trading of the MG with the main grid, and X_t^{MG} is a binary variable used to model this fact the MG cannot purchase/sell energy from/to the grid.

- The SLs' constraints: The amount of shifted up and down demand is limited as modeled in (7). Equation (8) models the energy balance of SLs where the total amount of shifted up and down demand in the operation time period are equal to each other.

$$0 \le P_t^{SL\_DN} \le \alpha P_t^L \ , \ 0 \le P_t^{SL\_UP} \le \alpha P_t^L \quad \forall t \tag{7}$$

$$\sum_t P_t^{SL\_DN} = \sum_t P_t^{SL\_UP} \quad \forall i \tag{8}$$

where α is the percentage of SLs.

- Constraints of the EES: The power charging and discharging of the EES are limited to the maximum power capacity of the EES as modeled in (9) and (10), respectively. The stored energy in the EES has the dynamic behavior as modeled in (11) and (12). Also, this energy is limited to its minimum and maximum values as modeled in (13).

$$0 \le P_t^{ch} \le P^{batt,max}(1 - X_t^{batt}) \quad \forall t \tag{9}$$

$$0 \le P_t^{dis} \le P^{batt,max} X_t^{batt} \quad \forall t \tag{10}$$

$$E_t^{batt} = E_{t-1}^{batt} + P_t^{ch} - P_t^{dis} \quad \forall t > 1 \tag{11}$$

$$E_t^{batt} = E^{batt,ini} + P_t^{ch} - P_t^{dis} \quad \forall t = 1 \tag{12}$$

$$E^{batt,min} \le E_t^{batt} \le E^{batt,max} \quad \forall t \tag{13}$$

where $P^{batt,max}$ is the maximum power of the EES, E_t^{batt} is energy stored in the EES,$E^{batt,min}$ and $E^{batt,max}$ are the minimum and maximum limitations of energy stored in the EES,$E^{batt,ini}$ is the initial stored energy in the EES, and X_t^{batt} is a binary variable used to model this fact that the EES cannot charge and discharge, simultaneously.

- DG's constraint: The power generation of DGs are limited to their maximum values as modeled in (14). Also, the ramp-up (RUP) and ramp-down (RDN) limitations of DGs are modeled in (15) and (16).

$$0 \le P_{k,t}^{\mathrm{DG}} \le \overline{P}_k^{DG} \tag{14}$$

$$P_{k,t}^{\mathrm{DG}} - P_{k,t-1}^{\mathrm{DG}} \le RUP_k^{DG} \tag{15}$$

$$P_{k,t-1}^{\mathrm{DG}} - P_{k,t}^{\mathrm{DG}} \le RDN_k^{DG} \tag{16}$$

where $\overline{P}_k^{DG}$ is the maximum capacity of DGs, and RUP_k^{DG},RDN_k^{DG} are RUP and RDN limitations of DGs, respectively. For $t = 1$, the amount of $P_{k,t-1}^{\mathrm{DG}}$ is 0.

- Constraints of the ENS: The amount of ENS in each time step is lower than or equal to the maximum acceptable of the ENS as modeled in (17).

$$E_t^{ENS} \leq \overline{E}_t^{\text{ENS}} , \ \overline{E}_t^{\text{ENS}} = \beta P_t^L \qquad \forall t \tag{17}$$

where β is the percentage of the demand than cannot be supplied and $\overline{E}_t^{\text{ENS}}$ is the maximum acceptable of energy not supplied in the MG.

- The constraint of pollution emission: The total pollution emission of the MG in its operation time is limited to its total emission production limitation (EM) as modeled in (18).

$$\sum_t \gamma^{Grid}(P_t^{\text{MG,in}} - P_t^{\text{MG,out}}) + \sum_t \sum_{k=1}^{K} \gamma_k^{DG} P_{k,t}^{DG} \leq \text{EM} \tag{18}$$

The solution approach used for this tri-objective function is described in the next section.

5 Weighted Goal Programming

One of the well-known methods to tackle multi-objective problems is Weighted Goal Programming (WGP) which was introduced by Charnes and Cooper (1977). This technique takes into account ideal goals for each objective function and tries to minimize total deviations from these goals concerning the importance degrees (weights) of objective functions. The ideal goal related to each objective may be identified by solving the model individually with that objective function, or it can be set according to the attitude of the decision-maker(s). The developed model based on WGP is given as follows:

$$\text{minimize } Z_{GP} = (\frac{w_1 d_1^+}{IV_1}) + (\frac{w_2 d_2^+}{IV_2}) + (\frac{w_3 d_3^+}{IV_3}) \tag{19}$$

$$OF^{cost} - d_1^+ + d_1^- = IV_1 \tag{20}$$

$$OF^{emission} - d_2^+ + d_2^- = IV_2 \tag{21}$$

$$OF^{social} - d_3^+ + d_3^- = IV_3 \tag{22}$$

$$d_1^+, d_1^-, d_2^+, d_2^-, d_3^+, d_3^- \geq 0 \tag{23}$$

Equations (4)-(17) where IV_1, IV_2, and IV_3 show the ideal goal values for the 1st, 2nd, and 3rd objective, respectively. The positive and negative deviations from the 1st, 2nd, and 3rd objective functions are denoted by (d_1^+, d_1^-), (d_2^+, d_2^-), and (d_3^+, d_3^-), respectively. Since all the objective functions are of minimization type, it is ideal to minimize the positive deviation from IV_1 for OF^{cost}, positive deviation from IV_2 for $OF^{emission}$ and positive deviation from IV_3 for OF^{social}. Finally, to create an objective function without unit for the proposed WGP model Z_{GP}, each deviation term is divided by its ideal goal value.

6 Computational Results

This section provides the computational results of the study to address the validation of the proposed mathematical model using a numerical example. The proposed mixed integer-linear programming (MILP) model is run via CPLEX solver/GAMS software. Tables 1, 2 and 3 represent the input information of the example including the values of the parameters. Moreover, Table 4 displays the ideal goal values for the objectives which were calculated by solving the single-objective model concerning each objective function. Table 5 shows the output results in terms of the objective functions and WGP variables. The reported run time is 2.4 s. Finally, Tables 6 and 7, and Figs 1 and 2 give the optimal amounts of the key variables.

It should be noted, we added a very small value to the 3rd term of the WGP objective function in order to prevent a division by zero.

Table 5 represents the optimal values of the objective functions such that the main variables take values as presented in Tables 6 and 7, and Figs. 1 and 2. It should be noted that $P_t^{\text{MG,out}} = 0$, which means that the amount of sold power to the market is 0. The output power of the DGs and the purchasing power of the MG from the grid are given in Tables 6 and 7, respectively. The MGO decides to purchase more

Table 1 Input information of the DGs

k	C_k^{DG} (\$/MWh)	γ_k^{DG} CO_2 emission (kg/MWh)	$\bar{P}_k^{DG}$ (MW)	RUP_k^{DG} (MW/h)	RDN_k^{DG} (MW/h)
1	18	650	1	0.6	0.6
2	20	500	1.5	0.8	0.8

Table 2 Input information of the indexless parameters

$E^{batt,min}$ (MWh)	$E^{batt,max}$ (MWh)	$P^{batt,max}$ (MW)	$E^{batt,ini}$ (MWh)	η^{dis}, η^{ch}	γ^{Grid} (kg/MWh)	$P^{MG,max}$ (MW)	α	β	EM (kg)	w_1	w_2	w_3
0.4	1.8	0.9	0.5	0.95	600	3	0.1	0.1	62000	0.40	0.35	0.25

Table 3 Input information of the t-related parameters

t	P_t^L	P_t^{PV}	C_t
1	3.475	0	1.92
2	3.565	0	1.03
3	3.61	0	0.55
4	3.7	0	0.47
5	3.79	0	0.12
6	3.925	0	2.1
7	4.1	0	28.96
8	4.15	0.023	39.16
9	4.18	0.14	39.9
10	4.23	0.23	39.48
11	4.28	0.33	39.36
12	4.35	0.34	37.32
13	4.55	0.48	31.42
14	4.65	0.426	31.9
15	4.53	0.31	29.3
16	4.48	0.2	26.8
17	4.42	0.047	26.55
18	4.45	0.08	27.66
19	4.52	0.023	37.01
20	4.58	0	38.26
21	4.66	0	39.71
22	4.75	0	39.91
23	4.65	0	36.76
24	4.25	0	14.41

Table 4 Amounts of ideal goals

IV_1	IV_2	IV_3
1605.637	49831.900	0

Table 5 Output results

Variables	OF^{cost}	OF^{Emission}	OF^{Social}	Z_{GP}	d_1^+	d_2^+	d_3^+
Value	1929.112	57130.746	0	0.132	323.475	7298.846	0

Table 6 The output power of DGs

k	*t*											
	1	2	3	4	5	6	7	8	9	10	11	12
1	0.023	0	0	0	0	0.400	1	1	1	1	1	1
2	0.8	0.921	1.5	1.07	1.5	1.426	1.5	1.5	1.5	1.5	1.5	1.5
k	*t*											
	13	14	15	16	17	18	19	20	21	22	23	24
1	1	1	1	1	1	1	1	1	1	1	1	0.4
2	1.5	1.5	1.5	1.5	1.5	1.5	1.5	1.5	1.5	1.5	1.5	1.275

Table 7 Purchased power from the market

t											
1	2	3	4	5	6	7	8	9	10	11	12
3	3	3	3	3	3	2.1	1.212	0.267	0.602	1.022	1.075
t											
13	14	15	16	17	18	19	20	21	22	23	24
1.64	1.259	2.173	3	3	2.332	1.545	1.622	1.219	0.92	1.685	3

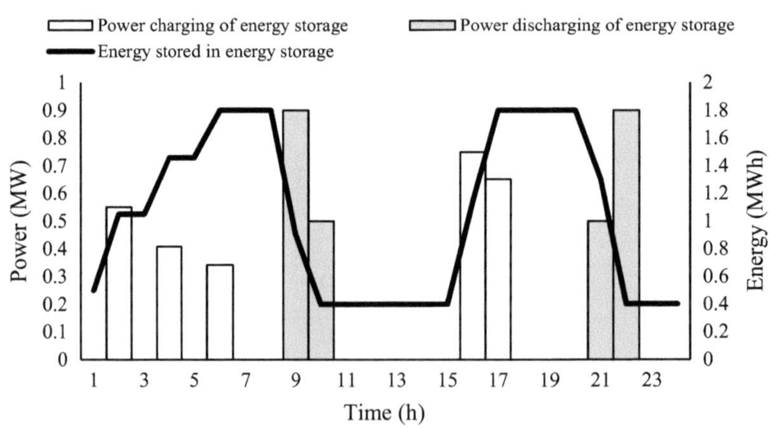

Fig. 1 The optimal scheduling of the EES in the operation time period

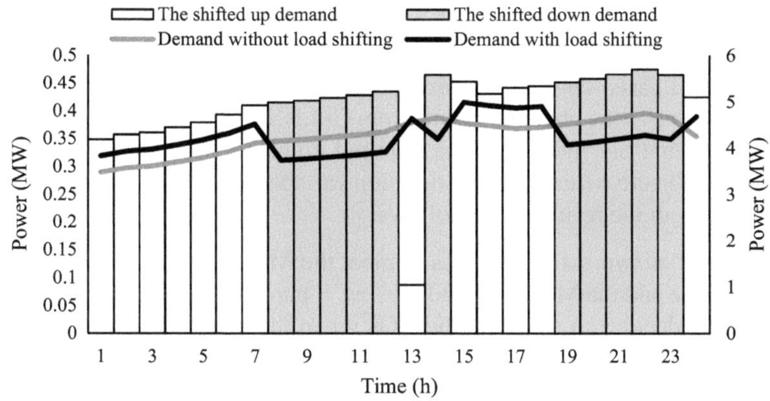

Fig. 2 The load shifting strategies of the MGO in the operation period

energy in hours 1–6 instead of dispatching DGs to meet its required demand. This decision is related to the low energy market prices in these hours in comparison with the generation cost of DGs. In other hours, since the demand of the MG and the energy market prices increase, the MGO schedules the DGs in their maximum power generation to meet demand with the minimum cost. From the viewpoint of the second objective function, producing energy by DG#2 leads to lower pollution emission in comparison with the purchasing power from the main grid. Although producing energy by DG#1 leads to higher pollution emission in comparison with the grid, its lower generation cost in comparison with energy market price assures the MGO to use the maximum capacity of DG#1. This decision is done regarding the proposed multi-objective model where the MGO's decisions are determined regarding a trade-off among objective functions.

The dynamic behavior of the EES is shown in Fig. 1. As shown in this figure, the MGO decides to charge the EES in hours 2, 4, 6, 16, and 17 where the energy market price is low. When the energy market price increases in hours 9, 10, 21, and 22, the EES is discharged. This behavior of EES can decrease the operation cost of the system. The optimal decisions of the MGO to shift the demand are shown in Fig. 2. The MGO decides to shift up the demand in hours 1–7, 13, 15–18 where the energy market price is low regarding which the demand of the MG in other hours with high energy market price increases. As shown in Fig. 2, the MG's demand profile changes after applying the demand shifting strategy by the MGO.

7 Conclusion

This study addressed the energy management problem of the MGs by developing a tri-objective MILP model. The objective functions were to concurrently minimize the operation cost, the pollution emission, and the energy not supplied. To deal with the tri-objectiveness of the proposed model, the WGP approach was implemented as one of the most efficient solution techniques. Moreover, a numerical example was considered based on the real-world situation to validate the proposed methodology. According to the obtained results from CPLEX solver/GAMS software, it was demonstrated that our model has the required efficiency to determine the optimal values of the objective functions and decision variables in a short run time. The main conclusions from the results are as follows:

- The MGO has two main strategies to meet the MG's demand. When the energy market price and the MG's demand are low, it purchases energy from the market. The MGO decides to schedule the DGs when the energy market price and the demand increase.
- The MGO uses the dynamic behavior of the EES and SLs to decrease the operation cost of the system. When the energy market price is low, the MGO charges the EESs and shifted up the demand to discharge the EES and shifted down the demand in other hours with high energy prices.

Based on the main limitations of the study, the following suggestions are given for future research directions:

(a) Applying uncertainty techniques including robust optimization, fuzzy programming, or stochastic optimal control to study the effect of uncertainty in the problem,
(b) Implementing other multi-objective solution methods such as ε-constraint to be compared with the proposed WGP,
(c) Considering other objective functions such as reliability maximization to address sustainable development in more detail.

References

Adefarati T, Bansal RC, Bettayeb M, Naidoo R (2021) Optimal energy management of a PV-WTG-BSS-DG microgrid system. Energy 217:119358. https://doi.org/10.1016/j.energy.2020.119358

Azimian M, Amir V, Javadi S (2020) Economic and environmental policy analysis for emission-neutral multi-carrier microgrid deployment. Appl Energy 277:115609. https://doi.org/10.1016/j.apenergy.2020.115609

Bahramara S, Golpîra H (2018) Robust optimization of micro-grids operation problem in the presence of electric vehicles. Sustain Cities Soc 37:388–395. https://doi.org/10.1016/j.scs.2017.11.039

Bahramara S, Mazza A, Chicco G, Shafie-khah M, Catalão JP (2020) Comprehensive review on the decision-making frameworks referring to the distribution network operation problem in the presence of distributed energy resources and microgrids. Int J Elect Power Energy Syst 115:105466

Bernardi E, Morato MM, Mendes PRC, Normey-Rico JE, Adam EJ (2021) Fault-tolerant energy management for an industrial microgrid: A compact optimization method. Int J Elect Power Energy Syst 124:106342. https://doi.org/10.1016/j.ijepes.2020.106342

Carducci F, Giovannelli A, Renzi M, Comodi G Improving flexibility of industrial microgrids through thermal storage and HVAC management strategies. Energy Procedia 142:2728–2733. https://doi.org/10.1016/j.egypro.2017.12.217

Charnes A, Cooper WW (1977) Goal programming and multiple objective optimizations: Part 1. Eur J Oper Res 1(1):39–54

Choobineh M, Mohagheghi S (2016) A multi-objective optimization framework for energy and asset management in an industrial Microgrid. J Clean Prod 139:1326–1338. https://doi.org/10.1016/j.jclepro.2016.08.138

Daneshvar M, Eskandari H, Sirous AB, Esmaeilzadeh R (2021) A novel techno-economic risk-averse strategy for optimal scheduling of renewable-based industrial microgrid. Sustain Cities Soc 102879 https://doi.org/10.1016/j.scs.2021.102879

Dey B, Shivam K, Bhattacharyya B (2019) Chapter 12–Energy management of various microgrid test systems using swarm evolutionary algorithms. in Chauhan RK, Chauhan K (eds) Distributed energy resources in microgrids. Academic Press, pp 323–341

Electricity final consumption by sector (2018) International Energy Agency. https://www.iea.org/data-and-statistics/?country=WORLD&fuel=Electricity%20and%20heat&indicator=ElecConsBySector

Golpîra H (2018) A novel multiple attribute decision making approach based on interval data using U2P-Miner algorithm. Data Knowl Eng 115:116–128

Golpîra H (2020) Smart energy-aware manufacturing plant scheduling under uncertainty: a risk-based multi-objective robust optimization approach. Energy 209:118385

Golpîra H, Khan SAR, Zhang Y (2018) Robust smart energy efficient production planning for a general job-shop manufacturing system under combined demand and supply uncertainty in the presence of grid-connected microgrid. J Clean Prod 202:649–665

Golpîra H, Bahramara S, Khan SAR, Zhang Y (2020) Robust bi-level risk-based optimal scheduling of microgrid operation against uncertainty. RAIRO-Oper Res 54(4):993–1012

Golpîra H, Khan SAR, Safaeipour S (2021a) A review of logistics internet-of-things: current trends and scope for future research. J Ind Inf Integ 100194

Golpîra H, Sadeghi H, Bahramara S (2021b) Electricity supply chain coordination: newsvendor model for optimal contract design. J Clean Prod 278:123368

Guo Q, Liang X, Xie D, Jermsittiparsert K (2021) Efficient integration of demand response and plug-in electrical vehicle in microgrid: Environmental and economic assessment. J Clean Prod 291:125581. https://doi.org/10.1016/j.jclepro.2020.125581

Lagrange A, de Simón-Martín M, González-Martínez A, Bracco S, Rosales-Asensio E (2020) Sustainable microgrids with energy storage as a means to increase power resilience in critical facilities: an application to a hospital. Int J Elect Power Energy Syst 119:105865. https://doi.org/10.1016/j.ijepes.2020.105865

Luo L, et al (2020) Optimal scheduling of a renewable based microgrid considering photovoltaic system and battery energy storage under uncertainty. J Energy Storage 28:101306. https://doi.org/10.1016/j.est.2020.101306

Naderi M, Bahramara S, Khayat Y, Bevrani H (2017) Optimal planning in a developing industrial microgrid with sensitive loads. Energy Rep 3:124–134. https://doi.org/10.1016/j.egyr.2017.08.004

Nakabi TA, Toivanen P (2021) Deep reinforcement learning for energy management in a microgrid with flexible demand. Sustain Energy Grids Netw 25:100413. https://doi.org/10.1016/j.segan.2020.100413

Restrepo M, Cañizares CA, Simpson-Porco JW, Su P, Taruc J (2021) Optimization-and rule-based energy management systems at the canadian renewable energy laboratory microgrid facility. Appl Energy 290:116760. https://doi.org/10.1016/j.apenergy.2021.116760

Zhang K, Li J, He Z, Yan W (2018) Microgrid energy dispatching for industrial zones with renewable generations and electric vehicles via stochastic optimization and learning. Phys A Stat Mech Appl 501:356–369. https://doi.org/10.1016/j.physa.2018.02.196

Trends in Sustainable Behaviour of Consumers in Eastern Europe and Sub-Saharan Africa: A Critical Discourse

Mirela Panait and Lukman Raimi

Abstract Across the globe, the issues of loss of biodiversity, industrial pollution, prolifreation of non-degradable plastics wastes, unhealthy mining activities, and climate change have imposed a new behavioural lifestyle on the individuals and the businesses in the society—a situation which academics and policymakers have tagged sustainabile behaviour. Scholars from the developed countries have written extensively on this important phenonomeon from multidisciplinary lenses. However, academic research on the sustainable behaviours of consumers are recently emerging in the Eastern Europe and Sub-Sharan Africa. This chapter critically discusses the trends in sustainable behaviour of consumers in Eastern Europe and Sub-Saharan Africa. Using a desk research technique, a qualitative research method as well as an interpretivist research paradigm, the authors provides a rich analytical tool for understanding the trends in sustainable behaviour of consumers from conceptual, theoretical and empirical studies in the two regions. In contextualising the trends, the paper found that in Eastern Europe and SSA the challenges of sustainability have forced human beings to adopt and adapt to green orientations and sustainable lifestyles called the sustainable behaviours. There have emerged in both contexts sustainable behaviour of consumers in the forms of green consumers, sustainable marketers, sustainable crop producers, sustainable agriculturists, green advocates, green policymakers, green politicians, conservationists, responsible investors or socially responsible investors, sustainable tourists and host of others. And the policy makers with support from hotels and green advocates, are influencing and reinventing consumers behaviour through campaigns and sensitisation initiatives aimed at reducing wasteful consumption, wasteful spending, irresponsible lifestyles, and habits.

M. Panait (✉)
Petroleum-Gas University of Ploiesti, Ploiesti, Romania
e-mail: mirela.matei@upg-ploiesti.ro

Institute of National Economy, Bucharest, Romania

L. Raimi
School of Business and Economics, Universiti Brunei Darussalam (UBD), Darussalam, Muara, Brunei
e-mail: lukman.raimi@ubd.edu.bh

C. J. Chiappetta Jabbour and S. A. R. Khan (eds.), *Sustainable Production and Consumption Systems*, Industrial Ecology,
https://doi.org/10.1007/978-981-16-4760-4_3

Keywords Sustainable behavior · Sustainable consumption · Green practices · Green facilities

1 Introduction

The loss of biodiversity, industrial pollution, proliferation of non-degradable plastics wastes, unhealthy mining activities, and climate change have generated many challenges for humanity (Zaman et al. 2012; Kruja 2013; Drăgoi et al. 2018; Gurtu and Arendt 2020; Hysa et al. 2020; Ali et al. 2021) and have imposed a new behavioural lifestyle on the individuals and the businesses in the society—a situation which academics and policymakers have tagged sustainable behaviour (SB). Some of the manifestations of the SB from environmental perspective include positive actions in favour of climate change, renewable energy, green consumption, carbon neutral, zero-carbon, eco-efficiency, energy efficiency, energy saving and waste recycling (Defra 2011, Harvey et al. 2014; Sima and Gheorghe 2015).

Gradually, all the efforts of the authorities, consumers, companies and other categories of stakeholders converge towards the same objective, namely low carbon economy or green economy. This ambitious goal generates not only challenges but also business opportunities that are exploited by companies, and paradoxes studied intensively, especially in developed countries, namely the green paradox or the Jevons paradox (Jean-Vasile 2015; Popescu and Nica 2015; Klumpp 2016). Intense economic growth in emerging countries generates, in certain periods, even a pollution of the environment, recent studies demonstrating the applicability of the Kuznets curve for the environment in different economies. (Armeanu et al. 2018; Hove and Tursoy 2019; Panait et al. 2019; Sharif et al. 2020).

Behavioral changes at the level of individual and company occur over time, the process being particularly complex and is influenced by many factors such as specific education, coercive measures by the authorities, the existence of a favorable legal and institutional framework, So, behavioural change may be self-induced or externally induced for personal or communal interests. Consumer behavior must be interpreted and analyzed in a social context, given that many of its actions can be explained by external factors such as belonging to a social group that generates certain acquisitions or manipulation by manufacturing companies or financial institutions that practice promotion campaigns fraudulent or abusive contractual clauses by which they try to use in their own interest the low level of education or the psychological resources that generate compulsive acquisitions In this context," challenge is to change our ways of thinking, feeling and acting, and to change society overall." (Figueroa-García et al. 2018, p.2).

Apart from the influence of social norms and public acceptability, scholars noted that imposition of prices (otherwise called price effects of desirable/sustainable behaviour) on the consumers by the government for discouraging the consumption of certain products/services especially consumption of alcohol, tobacco, and certain food has also resulted in human behavioural changes. Although the phenomenon of price effects on human behavioural changes often take years to drop certain lifestyles or breaking old habits (Lyons et al. 2009).

Scholars from the developed countries have written extensively on this important phenomenon from multidisciplinary lenses. However, academic research on the sustainable behaviours of consumers are recently emerging in the Eastern Europe and Sub-Sharan Africa. The paucity of comprehensive conceptual, theoretical, and empirical studies to bridge existing knowledge gap makes this study imperative and justifiable. The authors aim to add to the body of knowledge in this important and highly debated concept leveraging a qualitative research method.

This chapter critically discusses the trends in sustainable behaviour of consumers in emerging countries from Eastern Europe and Sub-Saharan Africa. The chapter focuses on sustainable consumption in emerging countries in Eastern Europe and Africa for the specific reasons.

- The emerging countries are in a process of economic development that involves changing consumption patterns on various levels such as energy consumption or food consumption.
- In these countries, banking and capital markets are not mature, consumers have a low level of financial education, and sometimes credit institutions take advantage of this situation through various methods such as deceptive promotional practices or abusive contractual clauses (Voica 2017; Panait et al. 2020),
- Foreign investors are interested in emerging markets given the business opportunities, the existence of consumers with purchasing power but also a labor force available at reasonable prices.
- These countries are also characterized by the intensification of the urbanization process which, in addition to the specific advantages, generates numerous threats to the protection of the environment, one of the biggest challenges being waste management and increasing of energy consumption (Feruni et al. 2020; Frone et al. 2020; Odugbesan and Rjoub 2020; Platon et al. 2020; Armeanu et al. 2021).

Two groups of countries were chosen for the analysis of sustainable consumption precisely to highlight regional characteristics but also to identify measures that can be taken in certain economies given the progress made in shaping sustainable consumption in other regions.

The emerging market concept is not an official one. Morgan Stanley Capital International proposed a list with emerging economies from different continents in order to calculate a stock exchange index (Annex 3.1). MSCI Emerging Markets Index was lauched in in December 1987 and covered only ten countries. Given the evolution of certain countries taking in account the changes of political regimes (like the fall of the comunism in Eastern Europe and of apartheid in South Africa) and national efforts toward development, the basket of the index has expanded considerably and since January, 2019, the index has 24 economies in the basket.

The economies from Eastern Europe are former communist countries that after 1990 initiated a process of transition to a market economy, a transition that in some cases ended, these countries being currently members of the European Union.

Membership of the European Union brings multiple benefits, at the legislative and institutional levels these countries being obliged to align with European standards. This fact generates positive consequences on the consumers' behavior, as they have at their disposal specific mechanisms and tools to have a responsible attitude towards the environment and society. In addition, EU efforts to promote sustainable development can be examples for other countries to follow."Transition from natural resources extraction, manufacturing and exporting to higher added value economic activity and domestic consumption" (Melas 2019). So, the process of economic growth specific to emerging countries generates not only positive effects such as increasing the living standards of consumers, strengthening financial markets, attracting foreign direct investment but also negative externalities such as environmental pollution or increasing volatility.

The specific issues that are focused in question forms are: (a) What is a sustainable behaviour and its associated factors? (b) What theory or policy made sustainable behaviour a front-burner issue in the contenporray times? (c) What are the trends in sustainable behvaiour in the Eastern Europe and Sub-Saharan Africa? (d) In contextualizing the trends in both continental regions, what are the theorerical and policy implications?

2 Conceptual Definition of Sustainable Behaviour

Multidisciplinary studies have described the influence of human behaviours as cyclical. At one end, human behaviours modulate personal states of people especially the habits, skills, and attitudes (Chemero 2013; Gardner and Rebar 2019), and at the other end, the personal states of the people influence human behaviours in return positively or negatively (Kaaronen and Strelkovsk 2020). It is also a truism that human behaviours are very influential in altering the environments where humans inhabit (Chemero 2013). The changing nature of human behaviour justified the assertion that it is a phenomenon that is socially constructed, learned and transmitted (Odling-Smee et al. 2013; Kaaronen and Strelkovsk 2020). In the face of existential issues and threats confronting the world's seven billion people, it became expedient to socially reconstruct human behaviour to adopt and adapt to sustainability principles, hence the idea of sustainable behaviour was conceptualised (McCool et al. 2013; Brandon et al. 2017; Kaaronen and Strelkovskii 2020). What is a sustainable behaviour?

The term sustainable behaviour (SB) refers to a set of planned, deliberate and responsive human actions aimed at conservation of natural environment and social resources for the purpose of protecting integrity of the society, and promoting positive psychological consequences such as satisfaction, self-efficacy, psychological wellbeing and restoration, happiness, and pleasure for individuals that adopt sustainable practices (Tapia-Fonllem et al. 2017). The concept of SB is also understood as a conscious and deliberate adoption by humans longstanding behavioural change in the forms of a pro-environmental habits, sustainable consumption and transport behaviours in an attempt to mitigate adverse effects of climate change and safeguard ecosystems (Kaaronen and Strelkovskii 2020).

Scholars stated that SB as a psychological concept is largely influenced by internal and external motivational factors. Humans by nature is positively motivated by pleasure, and negatively motivated by pain. It is believed that SB follows a normal curve with three motivational categories, namely: proactive behaviours, complacent behaviours and negative behaviour. Those with proactive behaviours and are self-initiated; those with complacent behaviours are forced to imbibe behavioural change by government or through motivational initiative; and those with negative behaviours are unmotivated and difficult to engage to adopt behavioural change to their lifestyles (Joyce Harvey and Norman 2004).

A number of research explained that humans as individuals adopt new sustainable behaviours for several factors. The study of Figueroa-García et al. (2018) proposed several factors that could favour the sustainable behavior - social pressure, government action, social environmental influence, demographic variables (age, gender, level of education), market conditions, education and information. The results of the scientific research reveals that environmental factor (like the influence of family and friends, cultural factors such as traditions), education and information and market conditions have a consistent influence of sustainable behaviour in Spain. The work of Koskela (2019) explained that there three major factors influencing SB in the society, namely: (a) demographic factors (such as age and gender), (b) internal factors (identify, values, nature connectedness, motivation and self-efficacy), (c) external factors (such as convenience of SB, social, cultural, political norms and practices).

However, specialists draw attention to the existence of a gap between consumers' attitudes and their behavior on the market (de Koning et al. 2016). Accroding with Arli et al. (2018)"consumers' positive attitudes about environmental issues do not necessarily translate into actual green purchase behaviour"(p. 389). The reasons why consumers, although aware of the existence and importance of environmentally friendly products or green products, do not buy them for various reasons such as economic constraints, product availability, perceived low performance, cynicism, confusion. The intention-action gap or green gap exist and the public authorities have the power to decrease it by specific measures. So, it is a challenge that must be managed by public authorities so that the intention to materialize and generate positive externalities both for producers and for society in general.

In some cases, consumers' reluctance towards environmentally friendly products is even justified given the greenwashing strategies practiced by some companies that want to fit the trend of CSR or transition to the low carbon economy manifested internationally, but their activity is devoid of substance (Vollero et al. 2011; Vollero et al. 2016).

3 Green Theory and Sustainable Behaviour

The concept of sustainable behaviour emerged to bring about sweeping and long-lasting behavioural changes to the reckless use of natural resources and wasteful consumption patterns in manner that threaten earth's capacity and ability to support humans and other species especially the flora and fauna (McKenzie-Mohr 2011). Therefore, the green theory provides a logical and consistent explanation for sustainable behaviour. In the field of international relations, the green theory with both normative and political arguments, asserts that individuals and corporations should be concerned with environmental and economic sustainability of their consumption and production activities in the society (Eckersley 2010). The theory advocates the deployment of green policies for regulation consumption and production with a view to preserving the ecosystem, rights of the people, ensuing social justice, responsible citizenship, good governance/democracy and eco-clean environment (Dunne et al. 2010; Eckersley 2010; Sima 2014; Gurtu et al. 2020). Consequently, there have emerged in the society different individuals and groups with green orientations promoting the ideals of SB in the forms of green consumers, sustainable finance, green advocates, responsible investors or socially responsible investors (SRI), sustainable supply value-chain among others (Khan and Yu 2019; Gurtu 2020; Kumar et al. 2020; Panait et al. 2020).

As good as sustainable behaviour directed at adopting a green lifestyle is, it is largely dependent on some enablers. The empirical study of Power, Beattie and McGuire (2017) identified eight factors as key enablers that motivate individuals to adopt sustainable behaviour and lifestyle. The eight enablers depicted in Fig. 1 below include access to green facilities, high standard green facilities, affordable green products relative to non-green alternatives, government initiatives that promote green lifestyles, green business and organizations, green technology, adequate information and education on green issues, and social support for green issues (Fig. 2).

4 Trends in Sustainable Behaviour in Eastern Europe

Eastern Europe is largely influenced by the trends in sustainable behaviour by the Western Europe as the consistent implementation of the principles of sustainable development (SD), taking in account that many countries from this region are EU members. European Union countries are promoters of responsible behavior for companies and consumers, there are numerous legislative and institutional initiatives that can be used as a model in countries outside the union. European officials are making remarkable efforts to establish the legal and institutional framework for citizens to manifest themselves as sustainable consumers on various levels such as energy or food consumption, or spending money to meet needs (Cătălin 2017; Cachero-Martínez 2020; Gurtu 2020; Zaman et al. 2020).The sustainable behaviour of consumers focuses on three basic sustainability levels, that is, environmental, economic, and social dimensions (Raszkowski and Bartniczak 2019).

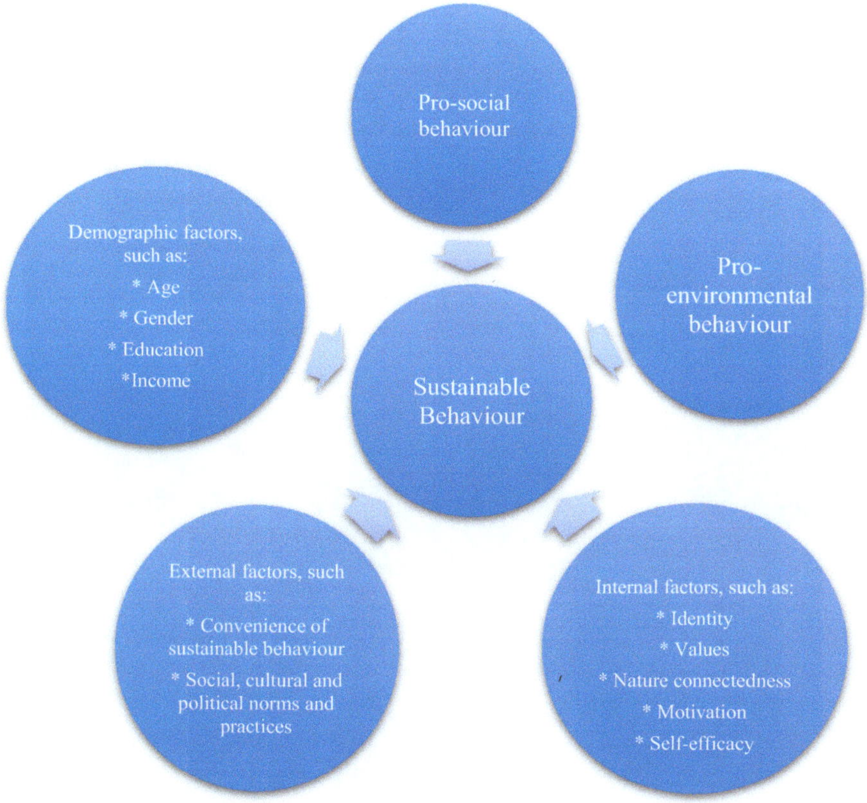

Fig. 1 Factors influencing sustainable behaviours. (*Source* Koskela 2019)

Western European countries are considered a setter trend for the rest of the continent given the level of development specific to these economies, which is reflected in a higher level of income but also a higher degree of awareness of the role of citizens and companies in the process. sustainable development complex.

Moreover, Folk (2021) noted that the trends in sustainable behaviour in Eastern Europe have taken the forms of increased focus on corporate social responsibility, consumer behaviour, environmental management, alternative renewable, smart technologies, and marine conservation (water and aquatic creatures). Specifically, corporate organizations in Eastern Europe have incorporated the act of giving back to the society and concerns for the environment as an integral part of business plans for the purpose of gaining competitive advantage and improve their revenue drives in the industry. Consumer behaviour is being reinvented through campaigns and sensitization initiatives aimed at reducing wasteful consumption, wasteful spending and irresponsible lifestyles habits. The role of consumers in the process of transition to the green economy is established by various European Union documents such as New consumer agenda"Strengthening consumer resilience for sustainable recover"

Fig. 2 Enablers to adopting sustainable behaviour

set up by European Union for period 2020–2025. This document is based on the 2012 Consumer Agenda (which expires in 2020) and the 2018 New Deal for Consumers (Fig. 3).

The efforts of the European authorities are reflected in the emergence of new consumption concepts and changing people's consumption patterns. These behavioral changes that adapt to the transition process of the green economy (such as repairs through community and social economy organizations actions like repair cafés and development of second-hand markets) are also potentiated by the COVID crisis which has demonstrated the importance of digital technologies in people's lives. Some of the changes produced during the pandemic such as the increase in the number and value of financial transactions through home bank or mobile bank, online purchase of food or accessing online streaming services at home will certainly become structural, the companies in the field being also in a process of adaptation and launch of specific products and services (EU 2020a).

Fig. 3 The main EU regulations on sustainable consumption

Reshaping consumer behavior in order to reduce the environmental footprint requires a complex information. In order to achieve long-term sustainable consumption, consumers should be provided with information on the lifespan of products, the availability of spare parts, repair manuals and specific services, with the aim of involving them in the process of promoting the circular economy. Therefore, consumers must have the right to repair damaged goods and to update their computer applications (EU 2020b).

Eastern Europe focusing eco-friendly environment by committing resources into water and reforestation timber and environment management. Additionally, governments, corporate bodies and individuals as consumers have embraced the use of solar, wind and other resources as renewable energy sources. Related to the above is the adoption of smart technologies in different spheres of life. The smart technologies allow connection of different electronic devices with sensors and internet for collection and and sharing of data with.

Another trend in sustainable behaviour is the growing concern for marine life conservation by the societies (Folk 2021). For instance, Russia as part of its commitment to the marine life conservation agenda of the Convention on International Trade in Endangered Species of Wild Fauna and Flora has taken stringent measures on illegal fishing and trading in endangered fish species such as sturgeon, salmon, crabs and cuboids, sea comb shrimps, sea-urchins and other species (Tishkov 2014).

It was reported that Czech Republic and Poland embraced an energy efficiency plan through the establishment of a coal-biomass plant funded by European Investment Bank with the intent of reducing CO_2 emissions by 24% including promotion of the use of small hydro, solar, and wind plants by the people and businesses. In Romania, a small thermal power plant that runs on sawdust from the local sawmills provide heating network for about 25% of the inhabitants of Vatra Dornei. This novel sustainable behaviour has reduced the import of liquid fuel (Mangalagiu and Jaeger 2016).

5 Trends in Sustainable Behaviour in Sub-Saharan Africa

Trends on sustainable behaviour of consumers in Sub-Saharan Africa (SSA) can be described as self-induced and externally-induced. In Nigeria, people are taking up sustainable behaviour otherwise called green behaviours in the forms of purchases organic food, use of energy-saving products, preference for green products because of environmental and health benefits (Ogiemwonyi et al. 2020a). In South Africa specifically, sustainable behaviour has become a fashionable lifestyle among students in the tertiary institutions because of high level of awareness about sustainability matters and the harmful effects of some products to the environment, Consequently, the students have embraced sustainable marketing because of their perceived sustainability benefits for humans and the society (Masocha 2018). The same trends manifest in other parts of the SSA.

Another dimension to sustainable behaviour that had been adopted in Africa for almost two decades is the concept of sustainable crop production (SCP). In the face of hunger and deprivation caused by rise in the demand for food that has outstripped the productive capacity of the land and current technology, the farmers embraced sustainable crop production by adopting a strategy of augmenting traditional soil enrichment practices. This sustainable strategy has functionally proven to economically feasible to many farmers and farming communities because it makes use of limited amounts of fertilizer and provides the best prospects for food security and sustainable agriculture in Africa's climatic zone (McCown et al. 1992). Moreover, SCP has been described in other African context as sustainable agriculture (SA)—a nuance that describes a food production behaviour with new opportunities that emphases the productive values of natural, social and human capital assets that increase food production at low financial cost (Pretty 1999).

Related to the trend above is conservation behaviour or movement. The movement is aimed at educating the consumers to be concerned about protecting and preserving endangered species of animal facing extinction in Africa. The electronic media is being used to spread environmental messages fostering sustainable behaviour and the benefits of environmental conservation (Pearson, Dorrian and Litchfield 2011). There is an extension of conservation education in Africa to sustainable tourism – a sustainable behaviour among the consumers/tourists with the intention to conserve and protect tourism sites through the enforcement of the sustainable tourism governance (STG) frameworks that emphases the need for trust, justice, social capital, power, and participation by the critical actors in sustainable tourism. The sustainable tourism intention has largely been successful through awareness creation, but the STG frameworks have been ineffective because the actors depend on mining and similar activities for sustenance, and mining-related activities affect sustainability (Siakwah, Musavengane and Leonard 2020). To strengthen sustainable tourism among customers in the SSA, the African and international certification programs was introduced to hotels and other players in the tourism industry with fabulous motivational incentives for the purpose of promoting and deepening sustainable behaviours and practices among customers as tourists and lodgers. The hotels are expected to encourage customers to be environmentally conscious, avoid wastage and conserve resources. In return, the hotels receive incentives such as marketing and promotion; interest free loans for acquiring new technologies; technical support and preferential inclusion in tour itineraries (Spenceley 2019).

As good as the emerging trends of sustainable behaviours in the SSA appear to look, a large scale adoption of sustainable behaviour is hampered by awareness, education, negative perception, distrust of green claims and contradictory behaviour of green consumers (Ogiemwonyi et al. 2020b).

For developing a good policy on sustainable behaviour in emerging contexts in Europe and SSA, analysts have recommended the use of psychological knowledge to complements the legislative approach and information-intensive campaigns, which are mostly used across the globe. This can best be implemented by investigating the potential behavioural barriers to sustainable behaviour or lifestyle changes before rolling out legislation or the implementation of a behavioural encouragement strategy (McKenzie-Mohr 2000a; Pearson et al. 2011). Based on the foregoing, best approach for future transition to a sustainable society is to leverage the concept of community-based social marketing, which essentially rely on the use of psychological knowledge regarding behaviour change to bring about the need behavioural changes in the society. The promoters of sustainable behaviour or sustainability issues deliberately identify social lifestyles and activities to be changed and the inherent barriers and then design a clear strategy for overcoming these barriers, using (McKenzie-Mohr 2000b).

6 Contextualizing the Trends in Both Regions

In contextualization the trends, the paper found that the challenges of sustainability in Eastern Europe and SSA have forced human beings as consumers to adopt and adopt to green orientations and sustainable lifestyles called the sustainable behaviours (SB). The trends in Eastern Europe and SSA have taken the forms of increased focus on corporate social responsibility, consumer behaviour, environmental management, alternative renewable, smart technologies, and marine conservation (water and aquatic creatures). The sustainable behaviors extend to deliberate preference for organic food, use of energy-saving products, preference for green products because of environmental and health benefits.

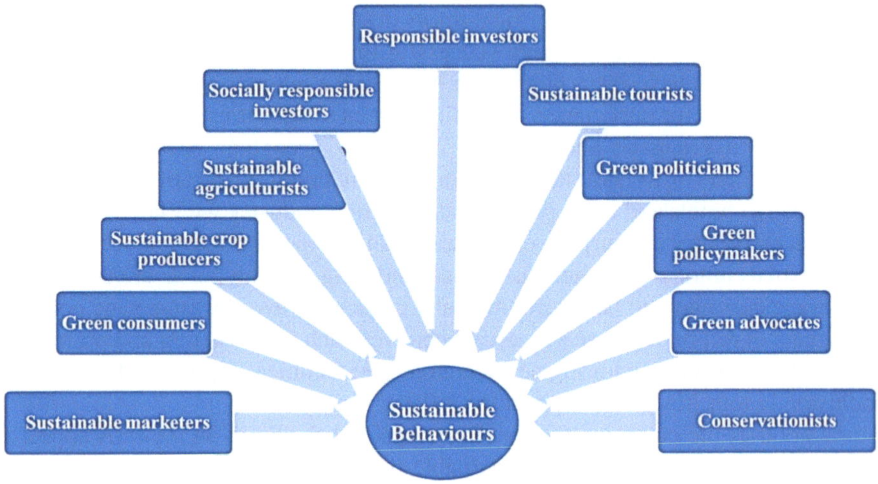

Fig. 4 Manifestations of sustainable behaviour of consumers

There have emerged in both contexts as shown in Fig. 4 below, sustainable behaviour of consumers in the forms of green consumers, sustainable marketers, sustainable crop producers, sustainable agriculturists, green advocates, green policy-makers, green politicians, conservationists, responsible investors or socially responsible investors, sustainable tourists and host of others. And the policy makers with support from green advocates, are influencing and reinventing consumers behaviour through campaigns and sensitization initiatives aimed at reducing wasteful consumption, wasteful spending, irresponsible lifestyles, and habits.

The practical and managerial implications of this paper are many and diverse, but the most important is that the policymakers in Eastern Europe and SSA need to develop green policies with robust measures and sustainable business models that would incentive green lifestyle. It is sensible that governments and non-governmental players continuously re-evaluate and re-assess the extent of actualization of sustainable production and consumption target (SDG 12) ensure that 'whatever it takes' sustainable behaviour should flourish in both contexts without any encumbrances. However, the qualitative insights and findings from this paper have some limitations. The purposive selecting of scholarly articles, texts and working documents is a limitation because the approach is viewed as non-random sampling and subjective. In this research, it was preferred because the intent was o survey articles that best present the trends in sustainable behaviour in Eastern Europe and SSA. Starting from the results of this scientific approach, the future research directions have in view the analysis of the sustainable behavior in the BRiCS and MINT countries taking into account the particularities of these countries. In spite of the above-discussed limitations, the paper is unique and distinct, as it opens the academic space further empirical and theoretical studies on this topical issue.

7 Conclusion, Implications and Policy Prescriptions

The paper sets out to critically discusses of the trends in sustainable behaviour of consumers in Eastern Europe and Sub-Saharan Africa using a desk research technique. At the end of the review of the trends in both contexts, it was found that the challenges of sustainability have forced human beings to adapt to challenges imposed by the necessity of low carbon economy establishment and adopt to green orientations and sustainable behaviours. There have emerged in both contexts sustainable behaviour of consumers in the forms of green consumers, sustainable marketers, sustainable crop producers, sustainable agriculturists, green advocates, green policymakers, green politicians, conservationists, responsible investors or socially responsible investors, sustainable tourists and host of others. And the policy makers with support from hotels and green advocates, are influencing and reinventing consumers behaviour through campaigns and sensitisation initiatives aimed at reducing wasteful consumption, wasteful spending, irresponsible lifestyles, and habits.

In view of the insights that have emerged from our policy cum conceptual paper, the following policy prescriptions are expedients and imperatives.

1. A combination of education, incentives and legislation is recommended as a potent approach for a sustainable behaviour in emerging contexts in Europe and SSA. It is therefore recommended that the policymakers in both contexts should complements the legislative approach and information-intensive campaigns with psychological knowledge.
2. Higher education plays an essential role in the process of promoting sustainable consumption through the competencies transmitted to graduates in the field of sustainable development, environmental protection, financial or energy education as future specialists and as consumers.
3. For long-term impact in line with the sustainable production and consumption (SDG 120 target, the policymakers in both Europe and SSA need to formulate and implement green growth policies that addresses critical issues on environmental degradation and climate change especially in the areas of carbon emissions, industrial pollution, plastics waste, land use, pesticide usage, fertilizer application, fossil fuel subsidies, cutting down of trees, and harmful mining activities.

4. The European Union can be considered a pole of excellence in the field of promoting sustainable consumption given the numerous legislative initiatives approved at regional level and transposed into the legislation of the member countries. Citizens' awareness of their involvement in promoting sustainable development through sustainable behavior has begun to bear fruit, despite the observation of a gap between their concrete intentions and actions. In addition, the steps taken in EU countries (some of them with the state of emerging markets) can be replicated in other countries, given the progress made at the level of this formation.
5. As being done in OECD countries, it is recommended that the policymakers consider funding public investment on sustainable behaviour and green lifestyles through the implementation of fiscal measures in three areas, namely: green taxation (environmentally-related taxes on emission), abolition of subsidies (phasing out subsidies granted to unsustainable projects and programmes) and adoption of green public procurement and expenditure (sustainable budgeting and spending). The first two areas provide sustainable financing and the third measure ensure focused utilization of the funds on green projects, portfolios and programmes.
6. Another important policy prescription is the need for periodic quality assessment of the degree of sustainable consumption. This serves two purposes: measuring progression in actualisation of SDG 12 and measurement of the extent of sustainable behaviourial change among people as individual and industrial consumers. Functionally, the OECD Green Growth Measurement Framework would be a useful and veritable tool for the policymakers the quest to assess if the level of economic growth encouraging sustainable behaviour and green lifestyle. The OECD Green Growth Measurement Framework has four dimensions, that are adaptable to all national contexts. These dimensions are: Environmental and resource productivity of the economy, Natural asset base, Environmental dimension of quality of life, and Economic opportunities and policy responses.
7. Despite the efforts made by the authorities at the institutional and legislative level, the transition to the low carbon economy is a complex process, instituted by certain challenges such as those generated by the Jenvons pardox or the green paradox. The process of economic growth is accompanied by numerous positive and negative externalities, the impact on the environment demonstrated the existence of the Kuznet curve and for emerging countries.

Annex 3.1 Additions to MSCI Emerging Markets and MSCI Frontier Markets Indexes

1988	• Argentina, Brazil, Chile, Greece, Jordan, Malaysia, Mexico, Philippines, Portugal, Thailand
1989	• Indonesia, Turkey
1992	• South Korea
1993	• Colombia, India, Pakistan, Peru, Sri Lanka, Venezuela
1995	• Israel, Poland, South Africa
1996	• China, Czech Republic, Hungary, Taiwan
1997	• Russia, Portugal
2001	• Egypt, Morocco
2006	• Bahrain, Kuwait, Oman, Yatar, UAE
2007	• Bulgaria, Croatia, Estonia, Kazakhstan, Kenya, Lebanon, Mauritius, Nigeria, Romania, Slovenia, Tunisia, Ukraine, Vietnam
2008	• Lithuania, Serbia
2009	• Bangladesh, Trinidad and Tobago
2012	• Saudi Arabia
2016	• West African Economic and Monetary Union (WAEMU)
2018	• China A
2019	• China A, Argentina, Sandi Arabia

Source Melas (2009).

"Country Introduced as Emerging Market/Frontier MarketTrinidad and Tobago, Ukraine and Bulgaria were removed from the MSCI Frontier Markets Index in 2001, 2015 and 2016, respectively.Venezuela was removed from the MSCI Emerging Markets Index in 2006. Saudi Arabia was re-introduced into the MSCI Domestic Indices as standalone index in 2012 and into the MSCI International indexes as standalone in 2015. In March 2018 the MSCI China A index was launched, offered in CNH and CNY versions. In May 2018 Large Cap China A shares were added to the MSCI China index, the MSCI Emerging Markets index and MSCI ACWI index at 5% of their FIF-adjusted market cap. The inclusion factor of China A will be raised to 20% in three steps in 2019". (Melas 2009. The Future of Emerging Markets. 30 Years On from the Launch of the MSCI Emerging Markets Index; MSCI).

References

Ali R, Nor SM, Naseem MA, Hasan M, Ahmad MI (2021) The integration of conventional equity indices with environmental, social, and governance indices: evidence from emerging economies. Sustain 13(2):676

Arli D, Tan LP, Tjiptono F, Yang L (2018) Exploring consumers' purchase intention towards green products in an emerging market: the role of consumers' perceived readiness. Int J Consum Stud 42(4):389–401

Armeanu DS, Joldes CC, Gherghina SC, Andrei JV (2021) Understanding the multidimensional linkages among renewable energy, pollution, economic growth and urbanization in contemporary economies: quantitative assessments across different income countries' groups. Renew Sustain Energy Rev 110818

Armeanu D, Vintilă G, Andrei JV, Gherghina ŞC, Drăgoi MC, Teodor C (2018) Exploring the link between environmental pollution and economic growth in EU-28 countries: is there an environmental Kuznets curve? PloS one 13(5): e0195708

Brandon PS, Lombardi P, Shen GQ (Eds) (2017) Future challenges in evaluating and managing sustainable development in the built environment. Wiley

Cachero-Martínez S (2020) Consumer behaviour towards organic products: the moderating role of environmental concern. J Risk Financ Manag 13(12):330

Cătălin VM (2016) Changes of EU industrial policy and its effects on romania's industrial paradigm. Industrija 44(4):197–210

Chemero A (2013) Radical embodied cognitive science. Rev Gen Psychol 17(2):145–150

de Koning JI, Ta TH, Crul MR, Wever R, Brezet JC (2016) GetGreen Vietnam: towards more sustainable behaviour among the urban middle class. J Clean Prod 134:178–190

DEFRA Department for Environment, Food and Rural Affairs (2011) Attitudes and knowledge relating to biodiversity and the natural environment 2007–2011: from the survey of public attitudes and behaviour towards the environment. DEFRA, London, UK

Drăgoi MC, Popescu MF, Andrei JV, Mieilă M (2018) Developments of the circular economy in Romania under the new sustainability paradigm. Econ Comput Econ Cybern Stud Res 52(2)

Dunne T, Kurki M, Smith S (eds) (2010) International relations theories: discipline and diversity, 2nd edn. Oxford University Press, Oxford

Eckersley R (2010:260) Green theory. In chapter 13, international relations theories disciplines and diversity second edition. Dunne T, Kurki M, Smith S (2010) New York: Oxford University Press

EU (2020a) Brussels COM(2020) 696 final communication from the commission to the European parliament and the council new consumer agenda strengthening consumer resilience for sustainable recovery.

EU (2020b) Brussels COM(2020) 98 final communication from the commission to the European parliament, the council, the European economic and social committee and the committee of the regions. a new circular economy action plan for a cleaner and more competitive Europe.

Feruni N, Hysa E, Panait M, Rădulescu IG, Brezoi A (2020) The impact of corruption, economic freedom and urbanization on economic development: western Balkans versus EU-27. Sustain 12(22):9743

Figueroa-García EC, García-Machado JJ, Perez-Bustamante Yabar DC (2018) Modeling the social factors that determine sustainable consumption behavior in the community of Madrid. Sustain 10(8): 2811

Folk E (2021) Top sustainability trends in Europe business, conservation and enviorenment. EcoMENA, Feb 16. https://www.ecomena.org/sustainability-in-europe/

Frone DF, Frone S, Platon V, Constantinescu A (2020) Green economy prerequisites of waste management. Sci Pap Ser-Manag, Econ Eng Agric Rural Dev 20(4):211–218

Gardner B, Rebar AL (2019) Habit formation and behavior change. In Oxford research encyclopedia of psychology

Gurtu A (Ed) (2020) Recent advancements in sustainable entrepreneurship and corporate social responsibility. IGI Global

Gurtu A, Arendt JD (2020) Packaging, business, and society. Int J Nonprofit Volunt Sect Mark 25(3): e1670

Gurtu A, Ranganathan S (2020) Sustainability from design to consumption. J Bus Econ Perspect 47, Spring/summer 2020, 68–90

Harvey J, Heidrich O, Cairns K (2014) Psychological factors to motivate sustainable behaviours. Proc Inst Civ Eng-Urban Des Plan 167(4):165–174

Hove S, Tursoy T (2019) An investigation of the environmental Kuznets curve in emerging economies. J Clean Prod 236: 117628.

Hysa E, Kruja A, Rehman NU, Laurenti R (2020) Circular economy innovation and environmental sustainability impact on economic growth: an integrated model for sustainable development. Sustain 12(12):4831

Jean-Vasile A (Ed) (2015) Green economic structures in modern business and society. IGI Global

Joyce S, Harvey J, Norman P (2004) Public perceptions of energy efficiency and recycling: how these can inform the communications process. In: Spitzer C, Schmocker U, Dang VN (eds) Probabilistic Safety Assessment and Management. Springer Verlag, London, UK, pp 3299–3304

Kaaronen RO, Strelkovskii N (2020) Cultural evolution of sustainable behaviors: pro-environmental tipping points in an agent-based model. One Earth 2(1):85–97

Khan SAR, Yu Z (2019) Strategic supply chain management. Springer International Publishing, AG

Klumpp M (2016) To green or not to green: A political, economic and social analysis for the past failure of green logistics. Sustain 8(5):441

Koskela IM (2019) Learning and sustainable development-a conceptual review of international literature on education for sustainable development. Report of finnish environment institute, 2019. https://helda.helsinki.fi/bitstream/handle/10138/306566/SYKEra_32_2019.pdf?sequence=1

Kumar A, Moktadir MA, Khan SAR, Garza-Reyes JA, Tyagi M, Kazançoğlu Y (2020) Behavioral factors on the adoption of sustainable supply chain practices. Resour Conserv Recycl 158: 104818. Kruja A (2013) Sustainable economic development, a necessity of the 21st century. MediterrEan J Soc Sci 4(10), 93

Lyons G, Musselwhite C, Pilkington P, Rayner G, Tapp A, Waygood EOD, Wiltshire P (2009) Individual behaviour change: evidence in transport and public health. https://uwe-repository.worktribe.com/preview/983024/avinerigoodwin_DfT_Individual_behaviour_change.pdf

Mangalagiu D, Jaeger C (2016) A third way out: the dilemma of eastern Europe. Solut J, Feb 22 Edition. https://thesolutionsjournal.com/2016/02/22/a-third-way-out-the-dilemma-of-eastern-europe/

Masocha R (2018) Sustainable marketing practices and sustainable consumer behaviour of tertiary students in South Africa. J Econ Behav Stud 10(3J): 248–257

McCool S, Butler R, Buckley R, Weaver D, Wheeller B (2013) Is concept of sustainability utopian: ideally perfect but impracticable? Tour Recreat Res 38(2):213–242

McCown RL, Keating BA, Probert ME, Jones RK (1992) Strategies for sustainable crop production in semi-arid Africa. Outlook Agric 21(1):21–31

Mckenzie-Mohr D (2000) New ways to promote proenvironmental behavior: promoting sustainable behavior: an introduction to community-based social marketing J Soc Issues 56(3): 543–554

McKenzie-Mohr D (2000) Fostering sustainable behavior through community-based social marketing. Am Psychol 55(5):531–537. https://doi.org/10.1037/0003-066X.55.5.531

Melas D (2019) The future of emerging markets. 30 years on from the launch of the MSCI emerging markets index MSCI

Odling-Smee FJ, Laland KN, Feldman MW (2013) Niche construction: the neglected process in evolution (MPB-37). Princeton University Press

Odugbesan JA, Rjoub H (2020) Relationship among economic growth, energy consumption, CO2 emission, and urbanization: evidence from MINT countries. SAGE Open 10(2):2158244020914648

Panait M, Radulescu I, Brezoi A (2020) Financial Markets-Under the Sign of CSR. Some Evidences Regarding Financial Education. LUMEN Proc 11:96–106

Panait M, Voica MC, Radulescu I (2019) Approaches regarding environmental Kuznets curve in the European Union from the perspective of sustainable development. Appl Ecol Environ Res 17(3):6801–6820

Pearson E, Dorrian J, Litchfield C (2011) Harnessing visual media in environmental education: increasing knowledge of orangutan conservation issues and facilitating sustainable behaviour through video presentations. Environ Educ Res 17(6):751–767

Platon V, Frone S, Constantinescu A, Jurist S (2020) Economic instruments for WEEE recycling in Romania. LUMEN Proc 14:509–523

Popescu GH, Nica E (2015) Global warming, climate policy, and the green paradox. In: Green economic structures in modern business and society pp. 20–38, IGI Global

Power N, Beattie G, McGuire L (2017) Mapping our underlying cognitions and emotions about good environmental behavior: Why we fail to act despite the best of intentions. Semiotica 2017(215):193–234

Pretty J (1999) Can sustainable agriculture feed Africa? New evidence on progress, processes and impacts. Environ Dev Sustain 1(3):253–274

Raszkowski A, Bartniczak B (2019) Sustainable development in the central and Eastern european countries (CEECs): challenges and opportunities. Sustainability 11(4):1180

Sharif A, Afshan S, Chrea S, Amel A, Khan SAR (2020) The role of tourism, transportation and globalization in testing environmental Kuznets curve in Malaysia: new insights from quantile ARDL approach. Environ Sci Pollut Res 27(20):25494–25509

Siakwah P, Musavengane R, Leonard L (2020) Tourism governance and attainment of the sustainable development goals in Africa. Tour Plan Dev 17(4):355–383

Sima V (2014) Green behaviour of the Romanian consumers. Econ Insights-Trends ChallS 66(3)

Sima V, Gheorghe IG (2015) Changing consumption patterns in green economy. In: Agricultural management strategies in a changing economy. pp. 186–212, IGI Global

Spenceley A (2019) Sustainable tourism certification in the African hotel sector. Tour Rev 74(2):179–193. https://doi.org/10.1108/TR-09-2017-0145

Tapia-Fonllem C, Corral-Verdugo V, Fraijo-Sing B (2017) Sustainable behavior and quality of life. In Handbook of environmental psychology and quality of life research. Springer, Cham, pp. 173–184

Tishkov A (2014) Conservation of biodiversity in Russia (executive summary of the 5th national report for the secretariat of the convention on biological diversity). Geogr Environ Sustain 7(3):125–136

Vasile V, Balan M (2008) Impact of greenhouse effect gases on climatic changes. measurement indicators and forecast models. Ann Univ Apulensis Ser Oecon 2(10): 1–19

Voica MC (2017) Financial inclusion as a tool for sustainable development. RomIan J Econ 44(1(53)): 121–129

Vollero A, Palazzo M, Siano A, Elving WJ (2016) Avoiding the greenwashing trap: between CSR communication and stakeholder engagement. Int J Innov Sustain Dev 10(2):120–140

Vollero A, Siano A, Palazzo M, Elving W (2011) Corporate communication and CSR; comparing Italian and Dutch energy companies on anti-greenwashing strategies. In: CSR communication conference 2011: the 1st international CSR communication conference.

Zaman G, Panait MC, Voica MC, Ene C (2020) Corporate social responsibility in the agri-food sector. In: Recent advancements in sustainable entrepreneurship and corporate social responsibility. pp. 37–68, IGI Global

Zaman G, Vasile V, Cristea A (2012) Oustanding aspects of sustainable development and competitiveness challenges for entrepreneurship in Romania. Procedia Econ Financ 3:12–17

Post COVID-19 Development of Sustainable Production and Consumption Systems

Muhammad Umar, Hafiz Muhammad Zia-ul-haq, Shahzad Ali, and Mohd Yusoff Yusliza

Abstract Since December 2019, the spread of COVID-19 has brought severe impacts on the world's socioeconomic system. After such unprecedented socioeconomic crisis, it is important to anticipate post COVID-19 situation and provide appropriate recommendations for the world to recover in the best possible manner. The COVID-19 pandemic has left priceless effects on human life and has severely disrupted the sustainable production and consumption system in the world. This situation of crisis is being viewed as a deepest global recession that is going to leave extensive effects on both society and economy. Even before the spread of COVID-19, world was facing several issues of social, ecological, and economic systems due to constant industrial development and globalization. Unsustainable pattern of production and consumption is considered to be major cause of such environmental and economic deteriorations in the world. Sustainability advocates have been struggling to explore various ways to nurture sustainable consumption transition. In light of this, the spread of COVID- 19 has acted as a catalyst toward prevailing issues, particularly encouraging authorities to take aggressive actions to ensure wider sustainability. Hence, the purpose of this chapter is to explore and discuss post-pandemic situation in the development of sustainable production and consumption system. Particularly, this chapter discusses multiple future scenarios in the context of sustainable production and consumption system. Similarly, it also discusses various socioeconomic responses to the COVID-19 pandemic. Further, it also provides an assessment of COVID-19 effects on different socioeconomic aspects and post pandemic recovery strategies of the world for sustainable production and consumption with a focus on Europe and China. Most importantly, this study contributes by discussing that how the production and consumption system would work sustainably in post-pandemic era. Particularly, it recommends implementing industry 4.0 in production and consumption system that promotes deploying

M. Umar (✉) · H. M. Zia-ul-haq · M. Y. Yusliza
Faculty of Business, Economics and Social Development, Universiti Malaysia Terengganu, Kuala Terengganu, Malaysia

S. Ali
The Superior College, Lahore, Pakistan

C. J. Chiappetta Jabbour and S. A. R. Khan (eds.), *Sustainable Production and Consumption Systems*, Industrial Ecology,
https://doi.org/10.1007/978-981-16-4760-4_4

digitization in the manufacturing firms to manage their operations. As implementation of industry 4.0 technologies could help in improving waste management, decreasing resource-intensive consumption, enhancing efficiency in supply chain management, and improving environmental performance, thus, ultimately achieving socioeconomic sustainability goals.

Keywords COVID-19 · Sustainable production · Sustainability · Sustainable consumption

1 Introduction

The recent Coronavirus disease (COVID-19), originated in china, has become a major threat to the human life. Turning into a pandemic, this disease has greatly disrupted the lives and sustainable livelihood of the people. According to World Health Organization (WHO), as of 18th February 2021, numbers reached 109,217,366 confirmed cases of COVID-19 and 2,413,912 corresponding deaths (WHO 2021). Its global expansion has created a great challenge in the contexts of both health and economics for the whole world. Now, it is more than one year, since its outbreak, it has tested the capability of the world to respond and survive. The COVID-19 has caused a complex crisis involving multiple interconnected social and economic dimensions. According to a preliminary investigation held by International Monetary Fund (IMF), this pandemic has caused a crisis of about US $2 trillion which is greater than the Great Depression of 1930.

Certainly, the effect of this pandemic on human life is priceless and has severely disrupted the sustainable production and consumption system in the world. This situation is being viewed as a deepest global recession that is going to leave extensive effects on both society and economy (World Bank 2020). Most of the economies in the world faced significant contraction in their gross domestic product (GDP). It is estimated that only in 2020 global GDP is reduced by 4.4% (IMF 2020) and world trade volume is reduced by 9.2% (WTO 2020). The implementation of social distancing and lockdowns caused many business shutdowns and massive job losses. For instance, airline operations have been severely reduced and many airlines end up completely shutting down. Similarly, the automobile industry has witnessed a substantial production drop. Most of the financial markets faced the worst fluctuations and corporations have struggled to maintain industrial production (Cohen 2020). Tourism industry was severely hurt as tourists cancelled their trips. Most of the commercial activities involving restaurants, markets, cinemas, exhibitions, concerts, and sports events etc. were ceased. Similarly, public transport was another major industry that was completely disturbed in many countries. The Organization of Petroleum Exporting Countries (OPEC) reduced oil production by 10% to control prices that were dropping less than US$25 per barrel (Wells et al. 2020). In short, this pandemic completely exposed the vulnerability of global health and economic systems. Companies compelled their workers to take time offs and it is estimated that only in Midwestern region of America there was unemployment of around 18% (Aaronson et al. 2020).

Considering these negative impacts, governments and authorities are strongly required to formulate effective post-pandemic policies in the context of sustainable production and consumption. The purpose of this chapter is to highlight and discuss post-pandemic scenario and transition perspectives in the development of sustainable production and consumption. Even before the COVID-19 epidemic, there was an evolving premise that the globalization of neoliberal economics was bursting between the several issues of social, ecological, and economic systems (Wells et al. 2020). Increasing levels of personal, government, and corporate debt and sluggish global economic growth were the main indicators of this failure. In the western world, the protectionists had started raising questions against the corporate logic of global strategy (Cuervo-Cazurra et al. 2020). Hence, the COVID- 19 pandemic may have acted as a catalyst toward some of these evolving issues, particularly encouraging to take aggressive actions to ensure wider sustainability.

Since last couple of decades, sustainability advocates have explored various ways to nurture sustainable consumption transition. Unsustainable pattern of production and consumption is considered major cause of environmental and economic deteriorations in the world (United Nations 1992, 2015). In order to ensure prosperity, authorities have been developing effective strategies to promote less resource-intensive forms of consumption. According to United Nations Development Program (UNDP), a prosperous nation can be best described as a "runaway consumption train" (UNDP 1998). Consistent with this description, various other economic organizations, such as European Commission, Organization for Economic Cooperation and Development, United States National Academy of Sciences, Royal Society of London, and Nordic Council have highlighted the importance of developing ways of sustainable production and consumption to tackle poverty and environmental issues. Recently, Paris Climate Agreement of 2015 also acknowledged the sustainable production and consumption as a key factor in targeting environmental and economic problems (Alfredsson et al. 2018).

Since the United Nations Conference of 1992 in Rio de Janeiro on Environment and Development, the strain of sustainable production and consumption has emerged on the global agenda in three defined phases. In the start, the attention was given to promote efficient and cleaner processes for the manufacturing sectors. Then, during 2000s emphasis moved towards the development of strategies to enhance awareness among households and consumers to help them in making effective decisions on green products. At last, after the global financial recession of 2008, the need was realized to bring systematic change to reduce consumption (Cohen 2019). In light of this background, it is strongly required to anticipate post COVID-19 scenario and transitions. While the current pandemic is being considered as an evolving economic recession, it is fit to admit that sustainability proponents have implicitly struggled to seek similar objectives in terms of sustainable production and consumption transition (Cohen 2020).

However, the COVID-19 has dramatically transformed the situation that is going to accelerate a substantial transition in the future. For instance, physical activities have shifted to online interfaces, now the individuals have become more adapted to virtual technology for delivering education, operating businesses, and conducting other broader range of social activities. The situation during COVID-19 suggests that the implementation of lockdowns and quarantine policies have created innovative forms of consumer demand. The higher the risk of infection, the harder it will become to reverse these transitions. At the end of the pandemic, maybe society will forget the ways to do things before COVID-19. Nonetheless, the governments and corporations will make positive efforts to inspire consumers to come out of the house and revert back to the normal. Similarly, central banks will be willingly lowering the policy rates to accelerate economic recovery. Somehow, this pandemic situation is an opportunity for both authorities and consumers to reduce intensive-resource based consumption. At the same time, narratives of social distancing to avoid viral transmissions will strengthen individual-based consumption patterns instead of shared consumption. Despite what actually happens in the post pandemic scenario, the authorities must devise strategies to make sure that the COVID-19 outbreak leads to a sustainable production and consumption transition. At the end, this would be an effective strategy to offset the adverse miseries caused by this pandemic.

2 Sustainable Production and Consumption Systems

The COVID-19 pandemic has brought a dynamic change around the globe, where millions of people suffered from economic vulnerabilities and many countries faced negative economic growth. However, these vulnerabilities came with some positive aspects in the context of environment change, for instance, world witnessed improvements in terms of pollution and climate change because of less economic activities during pandemic. In the post COVID-19 period, governments and authorities will make aggressive efforts to recover economic situation. Therefore, it is expected to see a substantial growth in both production and consumption activities after the end of COVID-19 pandemic. Increasing consumption demand for resource consumption will result in environmental degradation and resource depletion. In light of economic recovery, with everyone to consume more and more, it is really crucial to ensure sustainable patterns of development.

In the post COVID-19 time, the transition towards sustainable production and consumption can be a significant element in achieving sustainable development goals. Sustainable production and consumption refers to meeting the needs of everyone while consuming minimum resources. In this way, it contributes towards green economy and poverty alleviation. Sustainable production and consumption is very crucial for poor households who are more dependent on natural resources. Hence, it helps in improving the quality of livelihood by creating better employment opportunities and developing strategies for poverty reduction.

During the World Summit of Sustainable Development in 2002, sustainable production and consumption emerged as an important part of global agenda to promote sustainability. However, there has been a slow progress because of complexities in promoting sustainable production and corporate social responsibility. Government regulations and incentives could play a crucial role in targeting sustainability issues. Similarly, achieving sustainable production and consumption requires a comprehensive approach, where collaboration between various stakeholders is crucial to develop and implement effective policies. It is necessary to not only focus on production but also to encourage consumers to make sustainable decisions. Thus, it requires all the members of the society including individuals, educators, and authorities to play a crucial role in sustainable consumption transition. This section of the chapter covers explanation of sustainable production and consumption concepts and its importance in terms of sustainability. Whereas multiple transition scenarios are to be discussed in the next section to develop and implement supportive policies to achieve sustainable production and consumption development. Further, later sections present post-COVID-19 policy recommendations and transitions in the context of various major regions of the world.

3 Concept of Sustainable Production and Consumption

The basic idea behind the concept of sustainable production and consumption is to enhance quality of human life without causing resource exploitation and environment degradation. The concept has been defined in different ways. Recently it is defined by United Nations Environment Program as:

> "sustainable production and consumption is a holistic approach to reduce adverse environmental effects from consumption and production systems whereas enhancing quality of life for all". (UNEP 2011)

The concept of sustainable production and consumption stands on the notion of green growth resulting in green economy. The policy of green growth focuses on sustainable economic growth to promote low carbon and socially inclusive development. Whereas a green economy is described as one whose economic growth is based on investments that increase resource efficiency and decrease environmental hazards such as pollution and carbon emissions. Both green growth and green economy are viewed as key channels that can help in achieving goals of sustainable development.

Similarly, the concept of sustainable production and consumption also focuses on resource efficiency that refers to minimal resource consumption to achieve maximum output. Increasing resource productivity is the best way to achieve resource efficiency. Fuel efficient cars are the best example for resource efficiency strategies. Preservation of resources is important for the betterment of future generations. Thus, sustainable production centers on the best utilization of resources. Implying that all manufacturers must consider sustainable utilization of available resources during all stages of production. This resource efficiency is only achievable if individuals being consumers prefer sustainable products. Hence, resource efficiency is equally associated with both production and consumption concepts.

4 Importance for Sustainable Development

In the last century, the world has witnessed a remarkable progress for human kind. For instance, urbanization, economic outcomes, production, and population have increased around the globe. In most of the world these changes have contributed towards creating jobs, improving livelihood, reducing poverty, and economic development. However, this rapid development has come with a severe cost that also includes negative environmental stresses. There has been a dramatic growth in the consumption of natural resources (UNEP 2011). It was assumed that resources are in abundance and association between resource consumption and environmental hazards was ignored. Thus, high resource consumption caused various negative environmental outcomes such as air pollution, climate change, water shortage, and food insecurity. Further, it has also resulted in the scarcity of various natural resources which are considered crucial for the modern economic development (Weisz and Schandl 2008).

In addition to the environmental impacts, economic inequality has also increased between rich and poor in both developed and developing world. Moreover, COVID-19 pandemic has recently challenged the health and economic systems of the world, severely pushing socioeconomic conditions to the edge. Although the economic slowdown during COVID-19 has caused some positive environmental improvements. However, economies are strongly needed to be recovered and a high growth in production and consumption activities is expected in the near future. After the COVID-19, the current livelihood based on modern patterns of lifestyle will be requiring a large growth in resource consumption. Unfortunately, it would be very difficult for the world to afford such a large quantity of natural resource consumption while the earth's ecosystem will also be stretched towards danger.

Therefore, authorities are strongly suggested to focus on progressive transitions in the traditional system of production and consumption. Socioeconomic development, economic growth, and human wellbeing must be separated from huge natural resource consumption. In this regard, authorities are required to devise effective policies and strategies that help reforming economic system towards resource efficiency and sustainable production and consumption system. The twenty-first century is quite different form the twentieth century when resources were cheap and abundant. However, in the current economic scenario, resources are getting scarce and expensive that strongly requires to revise existing economic practices. In the future, economic development and prosperity will be determined by the efforts towards low carbon and green economy that is equally beneficial for both people and nations.

5 Post COVID-19 Transition Scenarios for Future Sustainability

The COVID-19 pandemic has already raised concerns regarding the role of government in any country in terms of socioeconomic sustainability. Developing a mechanism can be quite useful for enhancing the governance of post-COVID-19 transitions. Hence, this section of the chapter discusses different scenarios to analyze alternative future visions that will be helpful in understanding post-COVID-19 situation to come up with effective considerations. Two major characteristics of a society; economy and civil society, are considered to define and understand post-COVID-19 transitions. Where, economy includes economics, finance, business, and markets while civil society includes the state, governance, and regulations. The globalization is the most disrupted element during COVID-19 as trade and air travelling both have been severely hurt. Similarly, group based activities have also greatly suffered as compared to individual based businesses. It implies that as compared to small businesses in before COVID-19 situation, now large capital intensive businesses will be more vulnerable in the future. The given scenarios are more related to general economic conditions where existing assumptions regarding day to day business activities may no longer valid. Following the work of Wells et al. (2020) four transition scenarios are discussed below involving the roles of two major society characteristics which are civil society and economics.

6 Scenario 1: Business as Usual

The first scenario of business as usual is based on the idea that authorities will make aggressive efforts to revert economies and trade situation back to the normal. Though it might take some time to return back but it is possible. Hence, previous trends will recommence where changes can be predictable (Hickman and Banister 2007). Similarly, global trade will continue and economic growth will be determined by the material consumption. This scenario anticipates that a weak civil society together with strong economic system can lead back to business as usual situation where resource consumption, pollution, and carbon emissions will still prevail. It will be very hard to persuade drastic changes in the overall environment.

Technology and competition will result in environmental efficiencies but these efficiencies will be refuted by continuous increase in consumption. Business as usual is a controversial scenario as existing issues will still prevail. World Trade Organization (WTO) also expressed this scenario in April 2020 (Wells et al. 2020). This scenario is supported by neoclassical economics and assumes that globalization and its economic benefits are the best way forward for the future after COVID-19.

As authorities are already making efforts to protect economies, so, a return to previous routine is quite the possible. Recovery at best will take some years, however, ecological problems will continue to increase with the reversion of previous structures. Business as usual situation might improve environmental conditions in the short run which also happened at the time of global financial crisis of 2008, however, ecological burden will ultimately rise in the long run. Further, the aim to revert economies could also cause delay in environmental regulations and other green policies. So, the threat with this scenario is that without crucial regulations and measures, return to previous routine would accelerate sustainability crisis.

7 Scenario 2: Chaotic Transition

The second scenario refers to the situation where a weak civil society along with fragile economic system causes chaotic transitions. This scenario is consistent with the theoretical perspective of catastrophe which leads to lower levels of both consumption and wealth creation. Usually, this situation emerges after the war or political instability. There are many example in the history that shows catastrophic collapses from various reasons (Tainter 1988). Most of examples are from the times of war which resulted in large portion of population as refugees or migrants. Such situations lead authorities to make aggressive efforts for the recovery.

Before COVID-19, prospects for chaotic situation were limited to the nuclear war which could cause such devastation. Chaotic situations can also arise as result of conflicts among societies and their economies (Motesharrei et al. 2017). After the COVID-19, the chances of failure of global economic system is not that unlikely as there were many prevailing conflicts across the globe. Therefore, the COVID-19 pandemic could provide a spark to any potential conflict such as the tension between proponents of globalization and nationalist that can disrupt the entire socioeconomic system. In this chaotic scenario, given by Meadows et al. (1972), political and economic systems fall at large scale, thus, resulting in isolationism, and business collapse which would significantly reduce environmental problems. However, this chaotic scenario is highly troublesome with large-scale societal consequences around the globe.

8 Scenario 3: Managed De-Growth

The managed de-growth is the third scenario where purpose is to shrink an economy (Kallis 2011). It is the case where stronger civil society is aimed to deliberately reduce the economy. This radical vision is consistent with the theoretical viewpoint of de-growth (Kerschner et al. 2018). This scenario anticipates that economic growth is inconsistent with the environmental sustainability, therefore, landscape problems can only be tackled by refusing materialism and promoting de-consumption (Kallis 2019).

It implies that material restrictions, tight regulations, and economic de-growth would cause rapid decline in environmental issues. This scenario is a transition of production and consumption system with existing themes of shared work, less materialized lifestyle, green taxes, and reduction of private wealth. A shift towards de-consumption and economic de-growth would ultimately result in less resource consumption. In this scenario, it is difficult to create balance between need of growth in developing world and managed de-growth in developed world. Although, the possibility of de-growth and chaotic transitions are very less likely as compared to business as usual and managed transition scenarios, however, the magnitude and timing of the crisis can bring such transitions. The longer time period of crisis leads to a difficult reversion towards existing normal. Thus, the intensity of crisis and situation of specific country could determine the best way forward.

9 Scenario 4: Managed Transition

In post-COVID-19 time, a strong civil society along with economy and business still intact, would be able to pursue effective policies to restrain ecological issues. The COVID-19 pandemic could be a transition opportunity towards green economic system based on sustainable consumption (Wells et al. 2020). This scenario of managed transition is completely consistent with the theoretical perspectives of green growth and environment friendly economic system (Stoknes and Rockstrom 2018). The managed transition could be quite helpful to revise economies in achieving more sustainable production and consumption system. It is implied to develop a successful circular economy that requires minimal consumption of natural resources. Therefore, this transition could be the best possible way to successfully decouple economic prosperity from resource intensive consumption (Stoknes and Rockstrom 2018; Wells et al. 2020).

Managed transition scenario also accepts green policies brought by both private and public sectors. Similarly, it is also beneficial for the corporations that can bring effective skills, financial capital, and technological expertise. This scenario also supports governance and regulatory mechanism of globalization. Hence, managed transition is the possible application of "porter hypothesis" which indicates that strong environmental regulations are beneficial for the companies and ecological system (Wang et al. 2019). Thus, the COVID-19 pandemic could be a trigger opportunity to start the process of managed transition to bring positive changes in behavior and the society relative to existing usual practices. In the post- COVID-19 period, managed transition could be helpful in reducing ecological issues and establishing a new revised economic structure.

Following this scenario, reduced resource consumption, involvement of innovative technologies, business support policies, global collaboration, and strong regulations could lead towards a more sustainable system with less ecological issues. However, after COVID-19, this way forward requires a strong global agreement and cooperation towards new sustainable production and consumption system. Hence, transition towards sustainable production and consumption system could be the best post-COVID-19 strategy with twofold objectives on environmental and economic challenges.

10 Socioeconomic Response to COVID-19

The United Nation has initiated global agenda on sustainable development for 2030, which targets multiple socioeconomic issues such as poverty, climate change, and environmental degradation (Di Marco et al. 2020). The goal number 3 of this agenda states that "development of healthy life and promoting wellbeing for all ages". The COVID-19 outbreak has caused numerous impacts on the socioeconomic system of the world. Thus, opening new dimensions for human sustainability and production industry. Most of the manufacturing organizations are trying to understand adverse effects of COVID-19. Different responses to the pandemic indicate that there is a great need for the policymakers to develop a long term plan in order to ensure and achieve sustainable development at the end. Therefore, this sections discusses variations happened in response to COVID-19 outbreak. Mainly this section covers behavioral and socioeconomic changes resulted due to COVID-19. This would help to understand sustainability situation in the post COVID-19 time period.

Different restrictions and policies imposed by the authorities during COVID-19 have resulted in intense behavioral shifts. For instance, social distancing has discouraged people to involve in group based activities. Similarly, many organizations have adopted work from home policy to ensure separation and social distancing. On the positive side, world has witnessed new opportunities for the people to develop competencies to adopt telecommuting and virtual platforms which has resulted in a shift from resource intensive transportation. Many people find these new work settings quite attractive in terms of cost and convenience (Sarkis et al. 2020). It is expected that such practices may become more usual in the future as people develop their comfort levels with more convenient routines. With improvements in the quality of virtual platforms, there would be less physical travels and teleconferencing will become more normal work activity.

Societal concern about contagion diseases is expected to prevail, thus, public gatherings and large scale entertainment activities will probably reduce in future. This will ultimately result in reduction of mass consumption activities along with travelling. During COVID-19, world has faced shortages of many goods because of supply-chain inefficiencies. In addition to this, people were forced to stay in extended quarantines which have lead them to adapt themselves to survive with minimum products. This would also probably result in a behavioral change towards spending simple lives.

The COVID-19 pandemic has also raised questions on the globalized production system involving international shipments and complex value chain structures. The pandemic situation has exposed dependence on just in time delivery system. In future, localized smarter logistic systems based on internet of things would be probably more effective in such situations. For instance, local sourcing would be quite easier knowing the location of products and their components (Sarkis et al. 2020).

Social distancing has opened paths towards adoption of new advanced digital technologies. Decentralized manufacturing system could be implemented because of technologies such as cyber physical system and digital automation. This technology based system could contribute in reducing resource intensive production. For example, computer based machines would replace human workers to perform different tasks (Corkery and Gelles 2020). In addition, these machines or robots have benefits of being operated over longer distances from remote locations. Similarly, use of virtual reality for facility designing is another revolution in terms of sustainable production. In future, adoption of these technology based system would be quite useful in reducing resource intensive production.

During COVID-19, many countries faced issues in managing availability of medical supplies and transferring them to most needy locations. There were repeated incidents of misallocations and shortages of supplies with tragic consequences. However, the timely implementation of industry 4.0, technology based manufacturing could have helped in dealing with such complications. In future, technologies such as internet of things and block-chain would provide enhanced transparency and traceability in supply chain system. Such technologies would be beneficial in saving time, energy, and resources.

Implementation of lockdowns and curtailment of various routine activities have caused a major drop in global energy demand, thus, damaging the whole energy industry. For example, 19 energy companies were bankrupted only in USA during COVID-19 pandemic (Jiang et al. 2021). The electricity consumption also decreased which have significantly affected generation of fossil fuels. According to Watts and Ambrose (2020), it is very difficult for coal industry to recover after COVID-19 pandemic. On the other hand, renewable energy sources are less damaged and expected to replace fossil fuel sources in the future. Similarly, global oil sector has also witnessed a huge loss due to COVID-19 outbreak. The demand of oil has significantly decreased in commercial, industrial, and transportation consumption. In the mid of 2020, various oil companies in USA bankrupted when oil prices reached negative (Egan 2020). Geo politics played crucial role in causing oil crisis when oil producers try to push their competitors out of the market. In future, global oil demand will be decreased, thus causing delay in many ongoing projects (Siddique et al., 2021). However, low prices would be an opportunity for the oil importing countries to support their projects. In the post pandemic, long term low prices would not only reduce tradition energy resources but could also trigger a transition towards renewable energy alternatives. Thus, resulting in a shift from oil dependence to other commodities (Gould and Al-Saffar 2020). Considering this situation, it is really crucial for oil corporations to explore more sustainable and innovative solutions to set future strategies. Hence, authorities should also devise effective plans to ensure transitions toward renewable energy sources in order to avoid such calamities in the future.

On the positive side, the COVID-19 pandemic has resulted in substantial environmental improvements. Various satellite observations have reported cleaner air around the world (Newburger and Jeffery 2020). This shows that how far human efforts can achieve collectively by significantly reducing emissions. In the post-COVID-19, when governments would be making efforts for economic recoveries, there might have no sustainable environmental improvements. Nonetheless these short run improvements during COVID-19 would help people to understand the consequences of sustainable consumption and environment friendly behavior. However, in the post-COVID-19, environmental improvements cannot be achieved by imposing restrictive policies on the cost of higher socioeconomic losses. Therefore, authorities are strongly required to shift towards renewable energy sources and environment friendly transportation. As long term environmental improvements would have significant impacts on human life and health conditions.

11 Post COVID-19 Development of Sustainable Production and Consumption Systems in China

The Covid-19 pandemic is a global crisis having esoteric impact on all aspects of society and economy. The distress caused by this pandemic resulted in global recession, which is the largest in eight decades (World Bank 2020). Different organizations (e.g. World Trade Organization, International Monetary Fund) predict decline in global trade volume and GDP. According to UNCTAD, foreign direct investment declined from 30% in 2020 to 40% in 2021. The lockdown and social distancing policies halted many businesses and caused increase in unemployment across countries. Although, the economic indicators show improvement from the second quarter of 2020 as lockdowns were lifted (IMF 2020). Governments in different countries are trying to restore their economies and effectively control the spread of this pandemic.

As the pandemic and the lockdown strategies have disrupted the global value chain and production system through restrictions on goods and people's mobility. The question arises that: Will China be able to manage their sustainable production after this pandemic? Given China has world's second-largest GDP and is known as global manufacturing plant and have significant role in global production. What will be China's strategy to restructure their global production, consumption, and supply chain sustainably in the era of post-pandemic? This section provides a touch from various angles about the Post Covid-19 development in China regarding sustainable production and consumption.

12 The Effect of COVID-19 in China

The Covid-19 outbreak was started from Wuhan, Hubei Province, in China, where restrictions were imposed across the whole country, affecting the production and consumption in the country. Within two month, this outbreak spread across different countries. Hence, put territories under lockdown and become global pandemic (Euronews 2020).

During this pandemic, such type of global polices were needed, which could lessen the economic crisis and the covid-19 cases. To stop the spread of the disease, counter measures were taken which has suppressed the global production and consumption; in turn severely affected the China's cross border trade. During COVID-19 pandemic, restrictions were imposed on travelling, trade, and production activities to stop the spread of the virus in China. Due to these restrictions, in the first quarter of 2020, the outward FDI and imports and exports declined (Duan et al. 2020). Moreover, the lockdown circumstances have drastically affected the people's economic and social lives; however, the effect of this pandemic was positive on environmental performance. As the closure of business, factories, and restriction on traveling reduced the consumption which in turn decreased the level of CO2, PM 2.5, and PM10 in the atmosphere, hence, caused improvement in air quality index. Liu et al. (2020) also assessed the air quality index of China and finds that there is improvement in the air quality index after the lockdown policies. Moreover, the data issued by National Aeronautics and Space Administration (NASA) indicates that the pollution in Wuhan reduced up to 30%. Along with that, the social block for Covid-19 has motivated consumers to adopt leftover consumption strategy due to restricted movement and income losses, which brought a positive change in behaviour regarding wastage of food (Jribi et al. 2020).

The lockdown policies have led consumers to adopt online purchasing of food which has caused increase in inorganic waste (Zambrano-Monserrate et al. 2020). Researchers have demonstrated that increased knowledge among consumers regarding sustainable consumption has led the organization to adopt and strengthen their sustainable efforts (Baier et al. 2020). For efficient sustainable production and consumption recycling and separation of solid waste is crucial for communities. Moreover, the consumption of energy and economic growth have reduced, which resulted in improved environmental quality. Liu and Song (2020) have researched that producing maximum food with less usage of water is a crucial task, especially in those areas where there is weak ecosystem and shortage of water. Furthermore, it is recognized that the conflict between economic growth and environmental pollution has restricted the development in China (Hu/Zhang/Lin et al. 2014). Before Covid-19, natural resources and ecosystem are adversely affected due to industrialization and globalization in China. Thus, how to build a sustainable society by eliminating pressure on environment and resources is a great challenge for China. The unsustainable production and consumption patterns are the greatest reason for environmental deterioration (UN 1992). To decouple the economic growth from environmental degradation, the government of China takes the initiative of sustainable development. Many practices and strategies were conducted to shift toward sustainable production and consumption in China. Although many lessons learnt from this pandemic will be served as a model to deal with the future pandemic. Still, a new sustainable model is needed for sustainable production and consumption in the post-pandemic era, and that sustainable production and consumption policies must be global polices.

13 China's Responses to the Pandemic and Strategies for Post-pandemic Recovery

In the first quarter of 2020, the economy of China shrank to a drastic extent but it returned to exponential growth in the second and third quarter of the 2020 (NBSC 2020). China is among the few economies that recovered rapidly from this pandemic, and its V-shaped recovery provides an encouraging outlook for the countries across the globe. This section discusses how China responds to the pandemic for post-pandemic recovery strategies.

14 Digital Economy in the Post-pandemic Era

The advanced digital technologies have played a significant role in the prevention and control of COVID-19 pandemic as well as in resuming and increasing consumption. In the era of post-pandemic, the digital technologies have provide new products, new services and new business model in education, healthcare, pharmaceutical and smart-homing sectors and enable economic growth and improve circular economy in country. Digital technologies are vital drivers in promoting sustainable economic recovery and in reducing disruption in manufacturing activities. The development of new technologies such as industrial internet of things, and artificial intelligence are providing growth opportunities for digital economy. China has made institutional improvement, enables novel innovation and could able to manage sustainable production and consumption through adopting digital economy policies.

14.1 Digital Transformation and Innovation

The lockdown and quarantine polices have pushed industries and organizations to adopt digital technologies to handle and manage their production operations and working. The use of digital technologies has driven innovation in the business process as people could manage their tasks from home. McKinsey and Company (2020) discussed that this pandemic had accelerated the use of digital technologies across the globe. The adoption of digital technologies is highest in Asia, which is greater than the average increase of 2017–2020. Also, the level of adoption of digital technologies to provide digital product, is increased in healthcare, pharmaceutical and profession service sectors.

Digital technologies have played a crucial role in controlling COVID-19 pandemic by providing tracking, online checking, monitoring of patients, and enabled telemedicine, which has solved the problem of uneven medical resources through real-time consultation. For diagnostic treatment, intelligent robots were used which has improved the efficiency of working and reduce the burden on health care workers and reduce the chance of cross-infection. Moreover, digital technologies have also supported the social life through providing platform of online education and by enabling tourism in China through providing information about safer places and safer hotel with the help of online mobile application.

The trend towards digitalization has been increasing in China during this pandemic (Jiang 2020). The use of advanced digital technologies helped in stopping the spread of this virus as people could manage their work from home, buy things online. Also, the growing trend towards digitization helped in managing the delivery of products and medicines to different places. Jiang (2020) explained that the future digital economy in China will contain 5G, internet of things, blockchain technology, cloud computing and smart sensors related industries. Moreover, new technologies will bring innovation and promote growth in China. Ba and Bai (2020) stated that this pandemic had expedited the digital economy in China especially in financial sector. They also explained that digital technologies have accelerated innovation in finance and supply chain finance, which helped industries to resume their production and become the key in economic recovery. Furthermore, the researcher also demonstrated the importance of reforms and opening a financial sector in the post-pandemic era. Due to this pandemic, the reinforcement in automation enabled China focused more on boosting domestic strategy rather than export-oriented strategy and make investment to enhance domestic industrial chain through innovation. Also, the cyber physical system and increasing advances in digital automation has helped to manage decentralized production system. These capabilities of technologies manage production operations through social distancing and help in reducing energy from travelling resources. The best example in this regard is the use of Kiva robots in warehousing, which were replaced with human labour and provide an additional advantage, that these robots can be controlled through computer-controlled machines from long distances. Moreover, the concept of cyber physical system is linked with sustainable production and consumption as it provide virtual reality through built environments and help reduce energy reduction because of stepping down need for travel to and from work. Furthermore, linking virtual reality with robotics will be helpful in times of emergencies when contagion is a major concern. Along with that, timely deploying industry 4.0 and smart manufacturing will help in removing various bottlenecks and complications in supply chain. The digital technologies such as internet of things, smart sensors, and artificial intelligence provide transparency, traceability and real-time data and monitoring systems which enable timely placement of order, in turn, save time, money, energy, and resources and led industries towards sustainable production. Thus, the use of digital technologies will help to manage sustainable production and consumption in the post pandemic era.

15 Post COVID-19 Development of Sustainable Production and Consumption Systems in Europe

In the Global Economy, sustainable consumption and production are the main driving force behind economic development. According to the United Nations, around 1 trillion worth of food items is wasted every year by customers and retailers. The United Nations also identifies that inadequate manufacturing practices and consumption is the main reason for the food waste. Food value is recognized in the COVID-19 when most people have no access to the quality of foods and services. COVID-19 has become the blockade in achieving sustainable consumption and is perceived as the primary economic slowdown source. To respond to the challenges of COVID-19, the world moved towards the digital community to ensure a sustainable consumption pattern. In this regard, the European Commission in February 2020 presents strategies to reshape Europe into a digital world to overcome these problems (Hovsepyan 2020). Therefore, the European Union has introduced a digital system to achieve the next five-year goals. These digital development initiatives have three primary objectives: firstly, the implementation of technology that enables people to achieve wellbeing and uphold European values. Second, to ensure an eco-friendly environment and, thirdly, sustain a healthy society under a digital democratic system. To accomplish these goals, digital systems were implemented in Europe to reduce the aftershocks of COVID-19. States in the European Union adopted several diverse solutions like; building advanced information sources, smartphone applications, and digitalized public services for reliable and speedily connectivity across the national and international boundaries. It seems essential to decide whether the digital solution is fit for its purpose and help the European Union manage sustainable consumption and production in post-COVID situations?

16 The Effect of COVID-19 in Europe

All over the world, the COVID -19 effect being observed on socio-economic status. According to CBI (2020), pandemics in European countries had disrupted the supply of goods and services. Major economies in Europe (France, United Kingdom, Spain, and Italy) recorded 200,000 plus causalities at the beginning of this pandemic, and almost 27 countries in Europe have failed to respond adequately. Furthermore, Nature (2020) recorded that in the earlier days of COVID-19 pandemic, European union failed to manage the protective equipment to stop the wide spread of coronavirus.

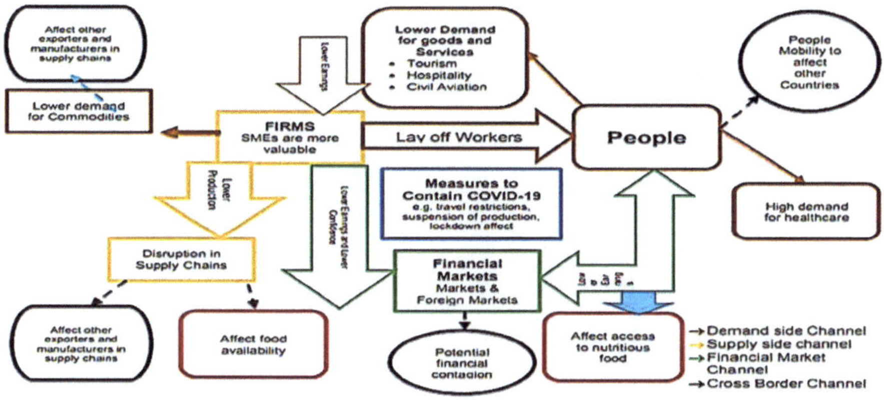

Fig. 1 Covid-19 impact on economy Meo et al. (2020)

Anyhow, countries across the globe allocated budgets to control this pandemic. Similarly, European Union also allocated $7.1 billion to manage this pandemic, but still, the European Union was helpless due to social distancing, quarantine, and lockdown policies. As Fig. 1 shows how COVID-19 has affected the economy, and explaining the four major interlinked parts of the economy, demand side channels, supply side channels, financial market channels and cross-border channels. During the COVID-19 pandemic, the implementation of lockdown policies forced all European economies to enter into hibernation. The manufacturing units and services sector were under lockdown, and the supply chain process was slashed. These all the factors cumulatively triggered the economy to fall by 8% as compared with 2019.

Different states in Europe have different impacts of COVID-19 pandemic; the Fig. 2 showed Italy and Croatia were the most affected states in Europe, and both countries' gross domestic product declined by 11% in 2020. Germany, Poland, and Hungary were also severely affected by COVID-19 pandemic, and their GDP declined around 4.6 – 7% just in 2020. Also in these regions, the income level of the

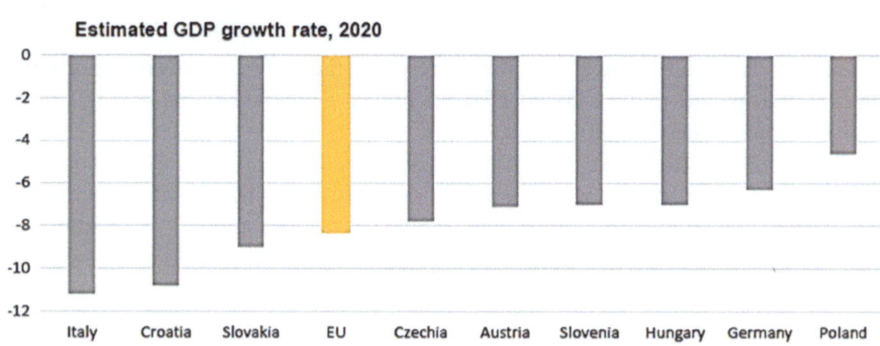

Fig. 2 Estimated GDP growth rate. *Source* EU Commission (2020)

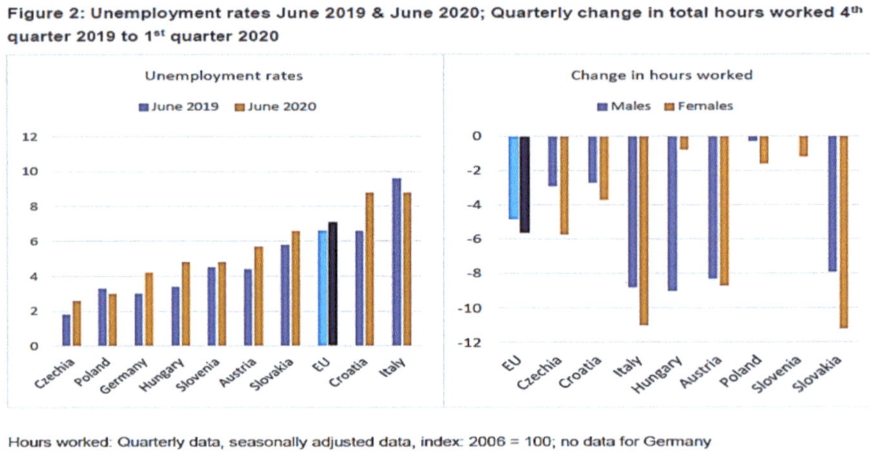

Fig. 3 Unemployment rate. *Source* Eurostat

people decreased up to 50% during the critical pandemic situation. Furthermore, the labor market in Europe was also victimized by pandemic, because of the reduction in the working hours; people were stuck in their home.

As earlier mentioned that labor was the primary victim of the COVID-19 pandemic, people faced job insecurity, while most of the companies fired their workers with the excuse of lockdown, and reduced working hours resulted in the reduction of wages, which ultimately affected the quality of life. The above Fig. 3 shows the unemployment rate of the European states, and comparative analysis of 2019 and 2020, which showed the great fluctuations in the employment rate. Due to the introduction of new health policies working hours were forcibly decreased to prevent the COVID-19 shocks, which resulted in an increase in unemployment.

Due to this pandemic, the industries in the region of Slovak and Poland were badly affected, while Croatia and Bratislava were less affected regions in Europe. On the other hand, regions like; Austrian, Czech, German, and Italy faced health crises owing to this pandemic. To measure the sensitivity of COVID-19 on regions, the index is constructed by Bohme and Besana by using the rigidity of lockdown, risk on macroeconomics. Conte et al. (2020) forecasted the GDP changes in the different states of Europe due to the COVID-19 pandemic, and concluded that the GDP of the whole Europe reduced by 20%.

The pandemic triggered the digital revolution and proved its effectiveness in keeping citizen's jobs and avoiding future social harms. The global pandemic has exacerbated the digital revolution and showed how important it is to minimize possible detrimental consequences, e.g., maintaining anyone active by teleworking (OECD 2020). A particular COVID-19 based result was a move toward teleworking. The ratio of people working from home before the COVID-19 pandemic was very low and not given much importance. But, pandemic circumstances had compelled people to work from home due to social distancing restriction and to avoid the spread of this virus. Also, the inclination to operate from home before the crisis was lower in the central European countries. In some European regions, especially in Poland, before this crisis, preference was given to work from home. In Poland, about 17% of the country's population operates from home. The tendency to work from home was less common in regions of Europe before this crisis.

The tourism industry was badly affected due to this pandemic across the globe. In Europe, the tourism sector was a substantial source of revenue and employment. Due to this pandemic, tourism in Europe had declined. According to the Euro-Stat, the most visited areas were Coastal regions in Croatia, Italy, and Germany, and were considered as the center for tourism. Due to this pandemic, the tourism activities were dropped which affected the employment opportunities and income of people in those areas. Comparing the tourism between April 2019 and May 2020, it was observed that tourism decreased significantly. The tourism activities declined from 84 to 98% in Croatia and 98% in Slovenia.

In Europe, due to this pandemic, the public transportation system collapsed because people were forced to stay at home. Freight transports deteriorated as direct consequences of travel limitations and reduced economic activities. Also, import and export are strongly declined in the European countries. In Slovakia, Czech Republic, and Hungary, the export dropped up to 15 to 20%. Le Quéré et al. (2020) mentioned that the reduction in the transport activities caused reduction in carbon emission from 11 to 25%.

During the COVID-19 pandemic, because of the associated contingency measures, the air quality across the globe improved. Due to lockdown and strict social distancing policies in Europe, the production in factories halted, and electricity consumption dropped by around 12 to 14%, which resulted in a decrease in the level of harmful emissions (such as CO_2, PM 2.5 and PM10). Hence, it improved the environmental performance in Europe. Despite this improvement in air quality and reduction in emissions, the negative effects of this pandemic resulted in reduction in recycling and an increase in household waste. Additionally, Zambrano-Monserrate et al. (2020) expected that with the increase in economic activities, the emissions will increase again in post-pandemic era, and the current positive environmental effects are only temporary.

17 Europe's Response to the Pandemic and Strategies for Post Pandemic Recovery

Monetary response is considered as fuel to come out from the pandemic situation. The European government provides the massive financial support for the households, companies, and for those whose earnings were at risk. Most of the European countries entered into the dangerous territory because they have no choice to support the failing banks, companies, and struggling households. In 2012, European commercial banks purchased the market bond to recover from monetary crisis (Grauwe 2020). During

COVID-19 pandemic, European commercial banks again adopted the same strategy and issued monetary grants to the member states to cover up the deficits in the crisis (Anderson et al. 2020). The inflationary rate under the pandemic situation arises, people faced difficulty to purchase goods and services especially food items in Europe (Grauwe 2020). European leadership was unable to take the discussion for the decisive common solution (Camous and Claeys 2020). German took the initiative on May 18, 2020, about the adoption of a digital economy to integrate the European Union. Furthermore, common fiscal policies were introduced for the Europe with shared responsibilities and announced 500 billion euros as the fund for the recovery of most affected regions. In this regard, European Commission also took the initiative in May 2020 and allocated 750 billion euros as the fund for recovery, in which 500 billion euro allocated to the grant and 250 billion euro allocated as backup loans for EU. In this context, authorities of Denmark, Sweden, Austria, Netherlands, are compromised and adopted the proposal of the European Commission and named the "Generation EU" fund for recovery. Earlier mentioned authorities, also allocated the 750 billion euro on the priority bases to reform their economies and achieve the financial targets. More than half of the fund almost 390 billion euros grants allocated for countries in the EU.

The wide spread of the COVID-19 exposed vulnerability in Europe, which compelled the European Union to make effective strategies to respond this pandemic. In this regard, a strong thrust needs to achieve the objective of European Green Deal which supports digitizing the business. To deal with this pandemic, and to control the unavoidable circumstance personal data were required, in this regard European Union used digital technology to manage it. In Austria, mobile application were used for tracking the people, which measure the distance between two mobile and helps the authorities to manage the Covid-19 cases. The European Union introduced the Electronic Cross Border Health Service for e-prescription and patient summaries to provide better health services through using industry 4.0 technologies. As a result of this pandemic, a large amount of private companies has already become digitized in Europe. The digital technologies has helped people to work, study and shop whenever they choose, as well as minimize travel which resulted in improved environmental sustainability in Europe.

In the background of the COVID-19 crisis, policymakers across the Europe have taken measures in place to cushion the global crisis and the decline in unemployment. According to the report of Amsterdam and Munich (2020) The European government has encouraged the private sector to work on developing new employment prospects and achieving a balanced pattern of demand and development thinking about economic policy and promoting investment. The pace towards digitalization and the emergence of new technologies had worked effectively during pandemic, can also work effectively during post-pandemic era regarding sustainable consumption and production.

18 Post COVID-19 Recommendations for the Policymakers

In the post-COVID-19 period, managing sustainable production and consumption would be a difficult task for the policymakers. In the light of current literature, this section provides some effective recommendations for the policymakers to achieve sustainable production and consumption goals after the COVID-19 pandemic.

18.1 Recommendations for Sustainable Production

Following the managed transition scenario, authorities and policymakers should devise national and global policies towards sustainable production and consumption system. Environmental burdens are more associated with the production side as compared to consumption. Regulation is the major tool available with the authorities to ensure sustainable production. Therefore, authorities should devise effective policies using taxes and subsidies to promote sustainable production system. Similarly, authorities can make corporate social responsibility compulsory for the companies. Authorities should consider providing incentives for the low carbon production in order to promote renewable energy usage. This would ultimately result in cleaner and sustainable production.

This pandemic situation could provide an opportunity to take a new start towards promoting sustainable production. For this, authorities could persuade producers to promote sustainable products by imposing efficiency standards on various products such as air conditioners, lighting, washers, refrigerators, and other household appliances. Additionally, authorities can also impose complete bans to remove less sustainable products from the market such as lead based paints. These product bans and energy standards could result in production of new innovative sustainable products such as efficient electronic appliances, and hybrid cars etc. Sustainable manufacturing could also be possibly applied in the health care sector around the globe.

Governments could also influence producers by adopting green procurement practices. In any country, governments are usually one of the largest buyers of goods. Many developed economies are already practicing green procurement. Green procurement includes variety of product categories such as recycles papers, efficient electronics, fuel efficient motor vehicles, information technology equipment, packaging, energy services etc.

Specifically, authorities should focus on industry 4.0 digital based manufacturing. Existing production facilities should be shifted to modern manufacturing using technologies such as Block-chain, 3D printing, and artificial intelligence etc. Similarly, authorities should also support adoption of modern information technology system to observe and analyze real production and consumption patterns, which could be quite beneficial for effectively managing production system in the post pandemic period.

Authorities should also make efforts to avoid system failure due to information gaps. In order to devise effective policies, there would be a great need to assure strong collaboration between different stakeholders such as authorities and industry managers to control and achieve targeted transition towards sustainable production. Additionally, mandatory labelling as government requirement is also effective strategy to ensure sustainable production. In this way, producers are bound to present social and environmental information of their product, for example, energy consumption and efficiency of the products.

18.2 Recommendations for Sustainable Consumption

The COVID-19 pandemic has caused severe deviations in the consumption pattern of people. Hence, policymakers and business managers must consider these variations in the demand for the necessary goods in the post-COVID-19 period.

Authorities should also make efforts to enhance awareness among consumers regarding social and environmental impacts of their day to day consumption. Label information could be a good strategy to educate public people about environment effects of the products. Similarly, governments can also include sustainable consumption practices as part of education.

Many consumers have adopted the behavior of piling important products during COVID-19 which has created problems in managing supply and demand for the producers. Authorities need to develop effective plans to enhance awareness against this behavior.

During the COVID-19 pandemic, the consumption of fossil fuels has been significantly reduced, thus, resulting in environmental improvements. Authorities and policymakers should devise effective policies to uphold such environmental improvements. This can be done by promoting renewable energy usage and its awareness among the public. Post- COVID-19 period would be the opportunity to accelerate and enhance sustainable consumption.

In order to target sustainable consumption, authorities can use taxes and subsidies which determines price of the product thus also influencing consumer behavior. For instance, fuel and vehicle taxes can be helpful in reducing pollution, and similarly, electricity and water charges can influence energy consumption. On the other hand, different incentives could be helpful in motivating consumers to go for sustainable consumption.

Social distancing and lockdown policies during the COVID-19 has increased the trend of online deliveries especially grocery and food items which has resulted in food wastage. Authorities should develop strategies for waste management as this trend would continue even after the end of COVID-19.

The COVID-19 has also enhanced the usage of social media in conducting all sorts of businesses. It is important for the business managers across the globe to analyze new behaviors and consumption trends. Policymakers should also devise proper strategies involving social media platforms to bring effective transition in consumption patterns.

18.3 Recommendations for Agriculture and Food

Agriculture, being an important area of sustainable production and consumption, is considered as source of opportunities for employment and gross domestic product of many economies. Furthermore, the agriculture sector is directly linked with food, and due to digitalization and advancement, stakeholders are reverting into agribusiness. Therefore, this sector also needs effective and efficient policies as it would play a crucial role to recover economies in the post-COVID-19 time.

Hence, there is a great need to develop and ensure sustainable national organic strategies targeting agriculture sector to achieve future sustainability. Specifically, authorities should devise agribusiness investment strategies in the COVID-19 time to enhance agriculture efficiency and quality of human life in the world. Similarly, land-use policies should also be developed for sustainable production and consumption to get the highest yield. Green finance is also required to be promoted to encourage sustainable production practices to achieve and ensure sustainability in agriculture and food sectors. Finally, authorities also need to improve their fiscal instrument, e.g. by reducing harmful subsidies, providing tax exemptions, using efficient energy resources, and promoting green entrepreneurship to achieve sustainability in agriculture and food sectors.

18.4 Recommendations for Waste Management

Sustainable development could not be achieved without implementing proper waste management. Social distancing and lockdown policies during the COVID-19 has increased the trend of online deliveries especially grocery and food items which has resulted in food wastage. Authorities should develop strategies for waste management as this trend would continue even after the end of COVID-19. Thus, there is a great need to establish values among the public for sustainable consumption and waste management. For this purpose, authorities should initiate awareness campaigns and develop policies on waste management education as such actions would directly affect public behavior, thus, ultimately enhancing the socio-economic and sustainability environment.

Authorities should also promote a regional value chain that integrates renewable use into efficient waste management by developing service areas, while creating revenue and jobs. Similarly, public–private collaborations should also be enhanced to encourage effective waste management services. So, authorities should establish networks that will allow collaboration, networking, sharing ideas, lessons learned, and opportunities for implementing effective sustainable waste management system. Authorities should also strengthen knowledge sharing with the private sector and establish regional networks for recycling and reuse.

In addition to this, authorities should develop a joint national information structure about sustainable waste management services for municipal institutions, public services, academic organizations, credential bodies, and government organizations. Also, promote, extend, and add measures to encourage clean energy such as biogas systems. Eco-labeling Scheme could also be quite useful in the future to enhance effective waste management. Further, sophisticated technologies could also be adopted by the authorities to track and monitor waste. Finally, authorities should develop and implement national electronic detection and surveillance network system for controlling hazardous waste.

19 Conclusion

The countries across globe face tragic time to skirmish covid-19 pandemic. This pandemic has caused disruption on larger scale and is considered as most fatal emergency across the globe as it has put every area of life under lockdown. Also, this pandemic has caused manufacturing firms to halt their production and to search for sustainable ways for smooth working of their supply and operations from customers and business perspective. To handle this and any future pandemic, the consumers as well as service and manufacturing firms need to be resilient and act sustainably.

This chapter provides an assessment of the impact of covid-19 pandemic on different aspects and post pandemic recovery strategies of the world for sustainable production and consumption with a focus on Europe, and China. Among major contributions, one of the observations is how the production and consumption system works sustainably in post-pandemic era. This study recommends deploying industry 4.0 in production and consumption system in the era of post pandemic as deploying digitization will help manufacturing firms to manage their operations virtually from far-off place which help reduce energy consumption because of stepping down need for travel to and from work in turn effectively manage sustainable production and consumption.

The pandemic situation has improved the environmental sustainability through reducing the carbon emission globally. On the other hand, people across the globe suffering health issues, job downsizing and financial burden because of production blockade and closure of sites. Taking these considerations into account, this chapter critically addresses aspects of sustainable production and consumption and directs the upcoming studies to identify the drivers and critical success factors to manage sustainable production and consumption in the context of post-pandemic and suggest policy frameworks for improving the resilience of production and operations processes.

Following are the implication for managers and practitioners: the production and consumption policies should be revised. As there are possibilities that the improved ecological performance of the nature will again start deteriorating during the pandemic in such a situation sustainable policies should be implemented in production firms. Along with that, industry 4.0 should be implemented for effectively managing the production system through less human efforts and through monitoring the systems from far off places. Moreover, deploying digital technologies provides online interface for conducting business and enable people to work virtually etc. All these things reduce the consumption of fuel which emits harmful emission and enable to retain the improved environmental sustainability. From consumption point of views, such type of policies should be made that could manage the supply chain disruption and ensure smooth flow of sustainable production. Furthermore, for sustainable consumption the supply chain managers need to revise their policies according to the requirement of the customers. All in all, this chapter recommends that for effectively managing the sustainable production and consumption there should be effective relationship between stakeholders such as manufacturer, NGOs, government, medical institutions and military agencies.

References

Aaronson D, Burkhardt H, Faberman J (2020) Potential jobs impacted by Covid-19 Accessed 10 Feb 2021. https://www.chicagofed.org/publications/blogs/chicago-fed-insights/2020/impacted-jobs

Alfredsson E, Bengtsson M, Brown HS, Isenhour C, Lorek S, Stevis D, Vergragt P (2018) Why achieving the Paris agreement requires reduced overall consumption and production. Sustain Sci Pract Policy 14(1): 1–5

Anderson RM, Heesterbeek H, Klinkenberg D, Hollingsworth TD (2020) How will country-based mitigation measures influence the course of the COVID-19 epidemic? The Lancet 395(10228):931–934

Ba S, Bai H (2020) Covid-19 pandemic as an accelerator of economic transition and financial innovation in China. J Chin Econ Bus Stud 18(4)

Ba S, Bai H (2020) Covid-19 pandemic as an accelerator of economic transition and financial innovation in China. J Chin Econ Bus Stud 18(4).

Baier D, Rausch TM, Wagner TF (2020) The drivers of sustainable apparel and sportswear consumption: a segmented kano perspective. Sustain 12(7). https://doi.org/10.3390/su12072788, 2788

Bergsen P, Billon-Galland A, Kundnani H, Ntousas V, Raines T (2020) Europe after coronavirus: the EU and a new political economy: royal institute of international affairs.

Camous A, Claeys G (2020) The evolution of European economic institutions during the COVID-19 crisis. Eur Policy Anal 6(2), 328–341

CBI (2020) Covid-19 makes a big change to the European consumption pattern. ministry of foreign affairs retrieved from https://www.cbi.eu/news/covid-19-makes-big-change-european-consumption-pattern

Chinese academy of information and communications technology (CAICT) (2020) White paper on the development of China's digital economy. Beijing.

Cohen MJ (2019) Introduction to the special section: innovative perspectives on systems of sustainable consumption and production. Sustain Sci Pract Policy 15(1): 104–110

Cohen MJ (2020) Does the COVID-19 outbreak mark the onset of a sustainable consumption transition? Sustain Sci Pract Policy 16(1): 1–3

Corkery M, Gelles D (2020) Robots welcome to take over, as pandemic accelerates automation. New York Times, 10

Cuervo-Cazurra A, Doz Y, Gaur A (2020) Skepticism of globalization and global strategy: increasing regulations and countervailing strategies. Glob Strateg J 10(1):3–31

Di Marco M, Baker ML, Daszak P, De Barro P, Eskew EA, Godde CM, Ferrier S (2020) Opinion: sustainable development must account for pandemic risk. Proc Natl Acad Sci 117(8):3888–3892

Duan W, Zhu S, Lai M (2020) The impact of COVID-19 on China's trade and outward FDI and related countermeasures. J Chin Econ Bus Stud 18(4):355–364
Economics, O. (2020). The economic impact of COVID-19 on Asia Pacific. Oxford Economics Ltd., vol 6, p 38-05
Egan M (2020) Oil prices turned negative. Hundreds of US oil companies could go bankrupt. CNN Business, New York, USA. https://edition.cnn.com/2020/04/20/business/oil-price-crash-bankruptcy/index.html
ESCAP (2020) Building a post covid-19 resilient economy. retrieved from https://www.unescap.org/blog/building-post-covid-19-resilient-economy
EU Commission (2020) GDP growth in the euro area and EU. News release Euro Indicators. Retrieved from https://ec.europa.eu/info/business-economy-euro/economic-performance-and-forecasts/economic-forecasts/summer-2021-economic-forecast_en
Euronews (2020) Coronavirus: half of humanity now on lockdown as 90 countries call for confinement. https://www.euronews.com/2020/04/02/coronavirus-in-europe-spain-s-death-toll-hits-10-000-after-record-950-new-deaths-in-24-hours
FAO (2010) Definition of sustainable diets. international scientific symposium: biodiversity and sustainable diets united against hunger: FAO Headquarters Rome, Italy
Gould T, Al-Saffar A (2020) Energy market turmoil deepens challenges for many major oil and gas exporters. IEA, Paris, France
Grauwe PD (2020) The need for monetary financing of corona budget deficits. Retrieved from https://www.intereconomics.eu/pdf-download/year/2020/number/3/article/the-need-for-monetary-financing-of-corona-budget-deficits.html
Harada K (2006) The green purchasing law, and promoting green procurement in Japan. paper presented at the presentation at ICPN workshop.
Hickman R, Banister D (2007) Looking over the horizon: transport and reduced CO2 emissions in the UK by 2030. Transp Policy 14(5):377–387
Hovsepyan E (2020) The European union's digital strategy and COVID-19. E-international relations. retrieved from https://www.e-ir.info/2020/05/19/the-european-unions-digital-strategy-and-covid-19/
Hu He, Zhang X, Lin L (2014) The interactions between China's economic growth, energy production and consumption and the related air emissions during 2000–2011. Ecol Ind 46:38–51
International Monetary Fund (2020) World economic outlook: a long and difficult ascent. https://www.imf.org/en/Publications/WEO/Issues/2020/09/30/world-economic-outlook-october-2020
International Monetary Fund (2020) World economic outlook: a long and difficult ascent. https://www.imf.org/en/Publications/WEO/Issues/2020/09/30/world-economic-outlook-october-2020
Jiang X (2017) Resource reorganization and the growth of the service industry in an interconnected society. Econ Res J 52(3):4–17
Jiang P, Van Fan Y, Klemeš JJ (2021) Impacts of COVID-19 on energy demand and consumption: challenges, lessons and emerging opportunities. Appl Energy 116441
Jiang X (2020) Digital Economy in the Post-pandemic Era. J Chin Econ Bus Stud 18:4
Jribi S, Ben Ismail H, Doggui D, Debbabi H (2020) COVID-19 virus outbreak lockdown: what impacts on household food wastage? Environment Dev Sustain 22 3939e3955. https://doi.org/10.1007/s10668-020-00740-y
Kallis G (2019) Socialism without growth. Capital Nat Social 30(2):189–206
Kemp PS, Froese R, Pauly D (2020) COVID-19 provides an opportunity to advance a sustainable UK fisheries policy in a post-Brexit brave new world. Mar Policy 120 104114
Kerschner C, W€achter P, Nierling L, Ehlers MH (2018) Degrowth and technology: towards feasible, viable, appropriate and convivial imaginaries J Clean Prod 197: 1619–1636
Le Quéré C, Jackson RB, Jones MW, Smith AJ, Abernethy S, Andrew RM, De-Gol AJ, Willis DR, Shan Y, Canadell JG, Friedlingstein P, Creutzig F, Peters GP (2020) Temporary reduction in daily global CO 2 emissions during the COVID-19 forced confinement. Nat Clim Change 10(7):647–653

Liu Q, Sha D, Liu W, Houser P, Zhang L, Hou R, Yang C (2020) Spatiotemporal patterns of COVID-19 impact on human activities and environment in mainland China using nighttime light and air quality data. Remote Sens 12(10): 1576
Liu Y, Song W (2020) Modelling crop yield, water consumption, and water use efficiency for sustainable agroecosystem management. J Clean Prod 253: 119940 https://doi.org/10.1016/j.jclepro.2019.119940.
McKinsey & Company (2020) How covid-19 has pushed companies over the technology tipping point and transformed business forever. https://www.mckinsey.com/business-functions/strategy-and-corporate-finance/our-insights/how-covid-19-has-pushed-companies-over-the-technology-tipping-point-and-transformed-business-forever
Meadows D, Meadows D, Randers J, Behrens W (1972) The limits to growth. Universe Books, New York
Meo MS, Sabir S, Chaudhry IS, Batool S, Farooq F (2020) Addressing the challenges of COVID-19 pandemic outbreak: Pakistan's preparations and response. J Public Aff, e2430
Mertens and van't Veer, P., Hiddink, G. J., Steijns, J. M., & Kuijsten, A. 2017 Mertens E, van't Veer P, Hiddink GJ, Steijns JM, Kuijsten A (2017) Operationalising the health aspects of sustainable diets: a review. Public Health Nutr 20(4): 739–757
Motesharrei S, Rivas J, Kalnay E, Asrar G, Busalacchi A, Cahalan R, Cane M et al (2017) Modeling sustainability: population, inequality, consumption, and bidirectional coupling of the earth and human systems. Natl Sci Rev 3(4):470–494
National bureau of statistics of China (2020) Economic growth of the first three quarters shifted from negative to positive. http://www.stats.gov.cn/english/PressRelease/202010/t20201019_ 1794616.html
Newburger E, Jeffery A (2020) Photos show impact of temporary air pollution drops across the world from coronavirus lockdown. NBC Universal Broadcast, Cable, Sports and News, Englewood Cliffs, New Jersey, United States
OECD (2020) Capacity for remote working can affect lockdown costs differently across places, OECD Policy Responses to Coronavirus (COVID-19). http://www.oecd.org/coronavirus/policy-responses/capacity-for-remote-working-can-affect-lockdown-costs-differently-across-places-0e85740e/
Payne CL, Scarborough P, Cobiac L (2016) Do low-carbon-emission diets lead to higher nutritional quality and positive health outcomes? a systematic review of the literature. Public Health Nutr 19(14):2654–2661
Sarkis J, Cohen MJ, Dewick P, Schröder P (2020) A brave new world: lessons from the COVID-19 pandemic for transitioning to sustainable supply and production. Resour Conserv Recycl 159: 104894
Siddique A, Shahzad A, Lawler J, Mahmoud KA, Lee DS, Ali N, Bilal M, Rasool K (2021) Unprecedented environmental and energy impacts and challenges of COVID-19 pandemic. Env Res 193:110443
Stoknes P, Rockstr€om J (2018) Redefining green growth within planetary boundaries energy. Res Soc Sci 44: 41–49
Tainter J (1988) The collapse of complex societies. Cambridge University Press, Cambridge
United Nations development program (UNDP) (1998) Human development report (1998) New York: UNDP. http://hdr.undp.org/sites/default/files/reports/ 259/hdr_1998_en_complete_nostats.pdf
United Nations development program (UNDP) (1998) Human development report (1998) New York: UNDP. Accessed 9 Mar 2020. http://hdr.undp.org/sites/default/files/reports/259/hdr_1998_en_complete_nostats.pdf
United Nations Environment Programme (2011) Resource efficiency: economics and outlook for asia and the pacific. CSIRO Publishing, Canberra
United Nations (1992) Agenda 21: Results of the world conference on environment and development. New York
United Nations (1992) Agenda 21. New York: United Nations. https://sustainabledevelopment.un.org/content/documents/Agenda21.pdf

United Nations (2015) Paris Agreement. New York: UN. https://unfccc.int/files/essential_back-ground/convention/application/pdf/english_paris_agreement.pdf
Vermeir I, Verbeke W (2006) Sustainable food consumption: exploring the consumer attitude–behavioural intention gap. J Agric Environ Ethics 19(2):169–194
Wang Y, Sun X, Guo X (2019) Environmental regulation and green productivity growth: empirical evidence on the porter hypothesis from OECD industrial sectors. Energy Policy 132:611–619
Watts J, Ambrose L (2020) Coal industry will never recover after coronavirus pandemic, say experts. https://www.theguardian.com/environment/2020/ma y/17/coal-industry-will-never-recover-after-coronavirus-pandemic-say-experts.
Weisz H, Schandl H (2008) Materials use across world regions. J Ind Ecol 12(5):629–636
Wells P, Abouarghoub W, Pettit S, Beresford A (2020) A socio-technical transitions perspective for assessing future sustainability following the COVID-19 pandemic. Sustain Sci Pract Policy 16(1): 29–36
Winter P (2020) The changing preferences of Asia-pacific consumers are creating opportunities for businesses that embrace an agile mindset. Retrieved from https://www.ey.com/en_gl/consumer-products-retail/how-asia-pacific-must-respond-to-the-new-post-covid-19-consumer
World Bank (2020) Global economic prospects. https://www.worldbank.org/en/publication/global-economic-prospects
World Bank (2020) Global economic prospects. https://www.worldbank.org/en/publication/global-economic-prospects
World Health Organization (2021) Accessed 18 Feb 2021. https://covid19.who.int/?gclid=CjwKCAiAmrOBBhA0EiwArn3mfIA5E6eg8mNBHmARaF8D8uT26EKKDlrNIhoX6zTmhg8-80H6blJYZxoC1y4QAvD_BwE
World Trade Organization (2020) Trade shows signs of rebound from covid-19, recovery still uncertain. https://www.wto.org/english/news_e/pres20_e/pr862_e.htm
www.imf.org/en/Publications/WEO/Issues/2020/09/30/world-economic-outlook-october-2020
Zambrano-Monserrate, M.A., Ruano, M.A., Sanchez-Alcalde L (2020) Indirect effects of COVID-19 on the environment. Sci Total Environ 728: 138813. https://doi.org/10.1016/j.scitotenv.2020.138813

Environmental Degradation and Sustainability Food Production and Waste Valorization: A Value Chain Analysis in Pakistan

Muhammad Asim Rafiq, Kiran Jameel, and Saifullah

Abstract In accordance with the global biodiversity creation of food waste and food processor's focus on improving food waste management strategies for the entire food supply chain (FSC), it is an essential need of time to conduct a comprehensive study. Consistent growth in urbanization, the transformation of lifestyles into modernization, and industrial re-evaluation will increase the use of natural resources. The current linear economic system, based on metal mining, used in manufacturing, consumes and ends up disposing of the product, which has caused environmental issues for the world, particularly developing nations such as Pakistan, as well as the loss of biodiversity, air soil water, and pollution, to endanger life on the Earth support system and to reinforce and maintain the soil in terms of land filling. Undoubtedly, this lack of resources is a major barrier to sustainable development and to addressing the growing needs of the community as well as increasing poverty. Developing countries such as Pakistan, which is part of a less developed world, where 29% of the population live below the poverty line, which can be translated into 55 million people. Community expectations have not been met due to problems such as rising unemployment, the worst working climate, poverty traps for social inequalities, widening inequalities, and intergenerational equity, and it is impossible for citizens to get three times a day's food. The primary cause is the food waste. The primary element in making our food structure more effective is to reduce losses across the entire FSC and reduce environmental degradation at the lowest level. There is a wealth of research available on the application of the Life Cycle Analysis (LCA) to various sectors such as agriculture, the production of different goods and their effects on environmental degradation, but rare studies are available in developing countries like Pakistan. This chapter aimed to cover all aspects of LCA, Life Cycle Costs (LCC), Environmental Life Cycle Costs (E-LCC), Social Life Cycle Costs (S-LCC), and their role in mitigating environmental degradation. Food waste with its three major classifications for e.g.; mega-trends, natural constraints; and root management were addressed in depth. Similarly, the effect of the climate on three main pillars of sustainability, such as economic, social, and environmental sustainability, has also been addressed. The proposed new LCA covers four closely linked methods which are the juxtaposition

---

M. A. Rafiq · K. Jameel (✉) · Saifullah
Hamdard Institute of Management Sciences, Karachi, Pakistan

C. J. Chiappetta Jabbour and S. A. R. Khan (eds.), *Sustainable Production and Consumption Systems*, Industrial Ecology,
https://doi.org/10.1007/978-981-16-4760-4_5

of substitutive product processes/services; the juxtaposition of a life-cycle substitute for a particular product or service; and the identification of life-saving changes in parts where maximum improvements can be produced. The application of LCA on various biological and non-biological food waste valorization process are elaborated. In the end detailed deliberation on current efforts made on LCA has been discussed in this chapter.

Keywords Environmental degradation · Food production · Waste valorization · Social life cycle costing · Life cycle costing

1 Introduction

The linear economic model centered on take, make and waste put the entire world in severe environmental degradation issues, a lack of natural resources to meet the need of the world population, a rise in poverty and hunger. While technological upgrades, the use of modern agricultural machinery, the efficient use of water supplies, the newest seeds, the use of insect-killing medicines and fertilizers have increased agricultural production, they still need to be integrated in a sustainable manner for coming generations. Food and agricultural organizations have considerable awareness of ensuring food security and nutrition to all human beings (FAO). Sadly, food produced across the world is not readily accessible to all citizens, as a large number of people dwell in underdeveloped areas (FAO 2009b). According to FAO, IFAD, and WFP 2014 estimates, around 805 million people were severely undernourished between 2012 and 2014. Reduction in starvation entails an integrated approach which include investment from private and public to enhance the agricultural productivity, optimal utilization of all available land, inputs service, latest technologies as well as the markets, steps to uphold the rustic development, protection for social, highly susceptible population and inclusion to strengthen the resilience from natural disasters and launching of nutrition program which encounter the micronutrient deficiencies in mothers and children under five years of age.

Presently, food safety and protection are under constant stress due to scarcity of natural resources, food inflationary pressures, unprecedented population growth, shifts in consumer lifestyles, macro-climate change, and food waste (FAO 2011a). As a precedent, the global population is projected to surpass about 9 billion in 2050, necessitating an expansion of 60–70% in food production. (Foresight 2011).Developing nations will play a key role in population growth relative to developed nations. The urbanization ratio will continue to increase, reaching 70% of the world's population (FAO 2009) Since 2007, the world's population is largely urban, although the United Nations World Urbanization Prospects (UNWUP) is projected to increase even more by 2025 (United Nations 2012).These unprecedented changes in urbanizations would certainly generate a market for livelihoods, shelter, and food supply. On the other hand, environmental change and a shortage of natural resources are restricting agricultural growth and food production. It requires a 70% increase in food production to feed the population of 9 billion, which will be very difficult to achieve (Hodges et al. 2010). The global degeneracy of natural resources, such as degradation in the supply of water, land, and biodiversity, would raise serious concerns to meet the demands of the future world.

Over the next 50 years, not only an increase in population but also an increase in urbanization is expected. This consistent growth in urbanization, the transformation of lifestyle into modernization, and industrial re-evaluation would expand the use of natural resources. The current linear economic system, based on metal extraction, used in manufacturing, consumes the product and ends up disposing of it, which caused the world, particularly developing nations like Pakistan, to environmental problems, as well as the loss of biodiversity, air soil water, and pollution, to jeopardize the life on Earth support system (Rockstrom et al. 2019; WWF 2014; Jackson 2009; Meadows et al. 2004) and to reinforce and maintain the soil in terms of land filling. Without a doubt, this lack of capital is a significant obstacle to sustainable development and to addressing the rising demands of the population as well as to increasing poverty. Developing countries such as Pakistan, which is part of a less developed country, where 29% of the population live below the poverty line, which can be translated into 55 million, people (International Monetary Fund IMF-2015). Public expectations are not being fulfilled amid fears such as massive unemployment, the worst workplace conditions, poverty trap social problems, widening disparities, and intergenerational equity (Banerjee et al. 2015; Prahalad and Ramaswamy 2004; Sen and Bhattacharya 2001). Financial issues like as complicated institutional ownership, risk in supply. inconsistent incentive systems and deregulated markets induce frequent economic and financial uncertainty for the entire economic system, as well as for individuals (Sachs 2015 and Jackson 2009).In order to tackle such incineration and other sustainability-related problems, such as environmental mortification, resulted in urbanization migration from rural to urban (Broad and Cavanagh 1993; Camp and Hofmann 1993; Field 1992; Homer-Dixon 1991, 1994; Jacobson 1989; Myers 1993, 1997; Suhrke 1994). In an effort to save the population from starvation, food production and consumption process need Improvement, especially in urban areas, which result in massive food waste. Gustafsson et al. (2011) It is estimated that 130 billion tons of edible materials are being discarded globally yearly, contributing to 33% of the world's food supply. About 20–30% of the ecosystem influenced by human food consumption and one-third of all food production waste or lost.

According to the study by Ceren et al. (2016), the worldwide ratio of avoidable food wastage is rising, with emissions (C_{H4}, 34 N_2O) of non-CO_2 Greenhouse Gas (GHS) rising more than three times between 1965 and 2010. The continued growth of farmland and prairies, which is determined by food waste, among other things, is the major underlying cause of environmental degradation and biodiversity loss (Donald and Evans 2006). As far as environmental change is concerned, it will not only affect agricultural production, but it also encompasses pollution from consumer transport of food purchased and consumer preparation can have a significant impact on overall outcomes. It is therefore essential to know at what stage of the food value chain food is being wasted. The cost of food waste worldwide would be around USD 01 trillion every 40 years. The cost of environmental destruction would be around USD

700 billion, in addition to the social cost of around USD 900 billion (Food Wastage Footprint–Impacts on Natural Resources. Summary Report, FAO, 2015) Food waste has adverse environmental, economic, and social consequences, and it's becoming a critical aspect of sustainable consumption. In addition to consumption, which is an obstacle to achieving sustainability goals, there are also a variety of explanations for producing food waste from producers and retailers. These factors are classified in three major classes.

(1) Natural constraints.
(2) Mega-trends; and.
(3) Management root.

Currently, various food waste management strategies have been put in place, such as the optimal use of energy, re-purpose, re-use, and recycling to reduce environmental impacts. The dream of a prosperous economy can be accomplished by introducing ingenious renewable energy technologies. These modern technologies have environmental and social advantages compared to the outdated disposal of models such as incineration and land-filling.

2 Environmental Degradation and Sustainable Food Production

Food waste is not only a developing country, but a global problem related to food safety and protection, and has other important sustainability aspects (FAO 2012) Food waste generation has several causes, some from food producers and retailers, which can be grouped into three major categories, i.e. natural constraints, mega-trends, and management (Mena et al. 2011). The food waste issue has a rigorous impact on three main pillars of sustainability.

2.1 Economic Impact

Food waste has an overall impact on the economy, such as significant crop losses during the processing or storage of food items. The economic impact can be calculated via the life-cycle costing technique (LCC).

2.2 Social Impact

Growing poverty in developing/low-income countries.

2.3 Environmental Impact

Nearly 14,000 million hectares of productive land (28% of the cultivated region of the earth). They are used annually to produce enough food which is wasted or destroyed. It is estimated that The carbon emissions of wasted food contributes to greenhouse gas (GHG) pollution through adding around 3.3 billion tones of CO_2.

2.4 Assessment of Environment Impact

The environmental effects of food waste can be measured using the following techniques:

- Life cycle Analysis
- Material flow Analysis
- Energy or Exergy Analysis

The term that is widely used in sustainable food production is organic farming, which Walter James said in his book "Look at the Land" in the 1940s. The topic of environmental concern was discussed in this book on chemical farming (Paull 2014). After this, sustainable performance has gained value in industry associated with food production (O'Brien 1999). Subsequently, sustainable performance has gained value in industry associated with food production (O'Brien 1999). While the idea of sustainable production was coined in the year 1992 at UNO in the Conference on Environment and Development, which is a core element of sustainable development, academics, scholars, and researchers have described sustainable production in a variety of ways. However, this term is well established by Lowell Center for Sustainable Production (1998) (LCSP).

- Generation of goods and services by this way that has non-polluting
- Protecting of natural resources and energy
- healthful and safe for communities, consumers as well as employees
- Economic viability
- Creatively rewarding to all such as society and people (Glavič and lukman 2007; Tseng et al. 2009).

Similarly, the American Public Health Association (APHA 2007) describes "one which delivers nutritious foodstuff to fulfill existing food needs and preserving healthy ecologies that can also deliver food for generations to come with least adverse ecological impacts. Sustainable food system also supports domestic production and distribution infrastructure and makes healthy food available, accessible, and affordable to everyone." It is inferred from the concept that sustainable production goals can be accomplished by preserving the current ecosystem in a healthy manner, minimizing waste that can protect the environment from degradation. The Sustainable Development Goals (SDGs) adopted by the Members of the United States cover both sustainable food production and consumption aspects within their established objectives. The Life Cycle Assessment (LCA) methods are well adapted for the promotion of these objectives. These techniques cover not only the stage of the food chain but also upstream, i.e., the processing of raw materials, animals, agriculture, fisheries, and fish farming, as well as the packaging of products. The downstream covers the distribution of goods, consumption, and disposal/waste also covers transport activities at all levels.

3 Life Cycle Analysis (LCA) of Food Waste an Overview

The Life Cycle Assessment (LCA) technique is applied to access the ecological impact of both food waste and urban solid waste. This technique is a revamped life-cost-cycle assessment (LCCA) approach in which the effect of 'cradle to cradle' has been measured. The modernist type of approach is the assessment of the life cycle (LCA). This is an approach for assessing the environmental effect of the commodity on its operation or service "from cradle to grave" from the stage of manufacture of the products to its final disposal as waste. The LCA methodology has been designed to resolve issues that have not been addressed by other legislative environmental assessment authorities. This approach has the potential to compare two or more alternative approaches in terms of their overall potential environmental effects on ecological sustainability.

The result of the LCA framework is pooled with engineering, social science, and economics to promote a universal approach to environmentally sound decision-making. The product life cycle (PLC) under the LCA covers all four phases, i.e., the period of introduction of the product, the phase of growth, the maturity stage of the product, and the decline phase of the product. The notion of LCA originated in USA at the end of the 1960s. For the first time in 1969, Midwest Institute the USA pursued the manufacturing technique of both plastic and glass bottles of Coca-Cola from resource to final disposal. Life Cycle Assessment (LCA) accompanied by Resource and Environmental Profile Analysis (REPA) as a base technique (Hunt et al. 1996) in late 1980, due to high national and international environmental degradation, awareness of the earth was steadily increased. The overall group is beginning to reflect on the outcome of the LCA. A significant number of LCA work has been facilitated in the growth of LCA theory. The first LCA International Seminar was held in 1990, organized by the Environmental Toxicology and Chemistry Society (SETAC). This seminar was the very first time that the idea of LCA was introduced. The first Online LCA Practical Guide was published in 1993, which makes it possible to conduct research on the LCA Methodology. The LCA Approach consists of four interrelated states (Figs. 1 and 2).

i. Identification of scope and goals
ii. The Analysis of inventory
iii. Assessment of impact and
iv. The interpretation (as reflected in figures)

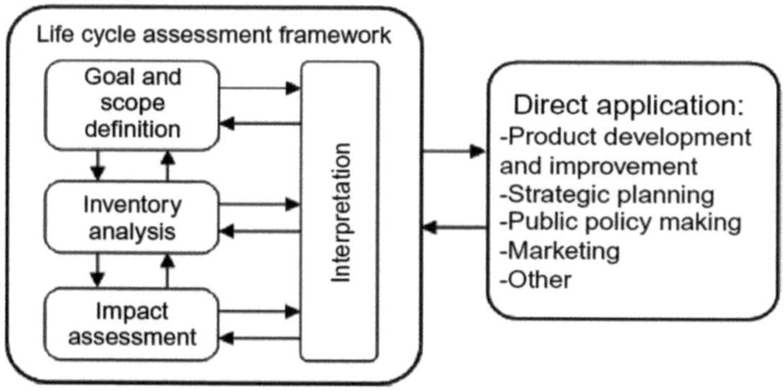

Fig. 1 Life cycle assessment framework (*Source* ISO 2006)

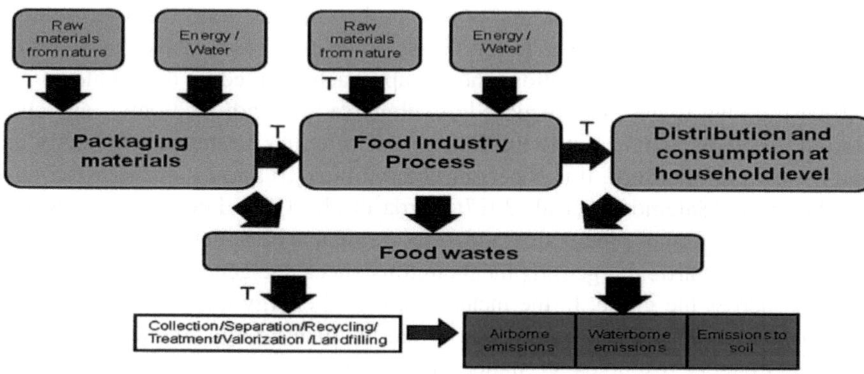

Subsystems, inputs and outputs considered in an LCA of a food product. T represents transportation. At household level, energy/water consumption may be needed as well.

Fig. 2 Sub-system in LCA of a product (*Source* ISO 2006)

3.1 Purpose of LCA

The key purpose of LCA approach consists of four interrelated process.

- Juxtaposition of substitute goods processes/ services
- Juxtaposition of substitute life cycle for specific product or services
- Identify improvement in the life of parts wherein maximum improvements can be done (Roy et al. 2009).

In 1997, the International Organization for Standardization fixed environmental management, the assessment of the life cycle with doctrines and structure (ISO 2006a).

4 LCA Analysis of Various Biological Food Waste Valorization Process

Globally, about 1.3 billion tons of food waste has been recorded annually by the Food and Agriculture Organization of the United Nations. This means that one-third of all food products are used for human consumption. This wastage includes the inclusion of agricultural, household, industrial, and commercial residues, while the compositional matrix of food waste depends mainly on the form and source of the waste (Xue et al. 2017).Food waste has a substantial loss of other natural resources, such as natural water, land, labor, and power. The Food and Agricultural Organization (FAO) has described food loss as food waste of quantity and quality during the overall supply chain process, captivating the place of production, post-harvesting, and the stage of processing. Food waste has a substantial loss of other natural resources, such as natural water, property, labor, and power. In direct terms, food waste resulting at the end of the food chain relates to food loss, which depends on the behavior of the consumer, the purchasing intention, and the approach adopted by the retailer.

Total food waste has been disposed of/submersed in landfills, humus, and ebullition in the wider range. While the European Union has strategies to directly use food waste as animal feed, this is becoming prohibited as a result of disease control apprehensions (Salemdeeb et al. 2017; Cerda et al. 2018). Food waste recycling Clinched production of other value-added food can also be both idyllic and end-use. Considering the particular geographical circumstances, the generation of food waste can differ across the globe. In the meantime, global energy utilization is enhanced more than twice during last twenty years (International Energy Agency IEA 2014). Traditional energy resources are under stress in major cities (Jacobson 2009), and food waste recycling has emerged as a viable and renewable resource that is proposed as a substitute for existing energy resources, as food waste has been identified as a third source out of 15 productive potential sources identified (Dobbs et al. 2011). As food waste recycling has been recognized as an additional source of energy that has also addressed the burning issue of food waste management, especially in big cities (Kiran et al. 2014; Ebner et al. 2014) In contemporary food waste management (FWM) framework, food waste recycling for valuable energy offers an admirable solution for food waste valorization. Anaerobic assimilation is a realistic solution for the reuse of food waste. We tried to develop a Table 1 containing the food wastage and their potential reuses (Fig. 3).

The anaerobic assimilation transforms food losses into valuable exergy in shape of biogas. This technique also has considerable merits due to involvement of least costs as well as low waste production (Morero et al. 2015). In addition; generation of biogas from food lost can be used in a various types for exergy utilization, such as electrical power and heat. This urban gas is a valued fuel that can be used for cooking purposes. On the other hand, biogas fuel can be used for vehicles (Lin et al. 2013) Due to a strong interest in transforming food waste into electricity, environmental assessment of this waste is of the utmost importance for policymakers to develop and design an excellent food waste management system. To estimate the recovery of

Table 1 Food wastage and their potential reuses

Vegetable or fruit		Usage	Main wastage	Wastage contains	Wastage uses	References
Fruits						
1	Apple	For juice 75%	Pomace	Hemicellulose = 9%	Food products	Shalini and Gupta (2010)
		By product 25%	Peel	Cellulose & lignin = 5%	Cattle feed	
					Pectin Extraction	
					Fuel purposes	
					Fiotransformation	
2	Berries	For jams & juices	Seeds	Fatty acid	Cultivation	Gorecka et al. (2010)
		For jellies		Oil content & triacylglycerol	Snacks & sweets preparation	
3	Citrus fruits	For juices	Seeds	Citrus fiber	Orange fiber	Bampidis and Robinson (2006)
		By-product	Peels	Orange juice wastewater		
			Pulp		Cattle feed	
4	Exotic fruits		Seeds	Pectin source, a dietary fiber	Amino acids; phenolic compounds	Ayala-Zavala et al. (2010)
			Peels		Bioactive compounds	
			Flesh			
Vegetables						
1	Potatoes	Food processing	Peels	Edible vegetable	Cattle feed	Cetkove et al. (2012)
2	Tomatoes	Food processing	Peels	Fiber and macronutrients		

food lost from an environmental point of view, the LCA is an analytical and scientific tool for connecting two or more alternatives in the context of potential environmental impacts (Cherubini and Strømman 2011).

5 LCA Analysis Of Various No Biological Food Waste Valorization Process

Through the LCA process, the sustainability goal of environmental degradation can be accomplished by reducing greenhouse gas (GHG) emissions, but liquid residues with a high organic content and a high quantity of solid waste and sludge are still a problem requiring the implementation of a non-biological solution. Animal fat and vegetable oil contained diesel fuel components known as biodiesel derived from alkaline esters (CnH2n + 1) produced from the dedicated crops of edible resources

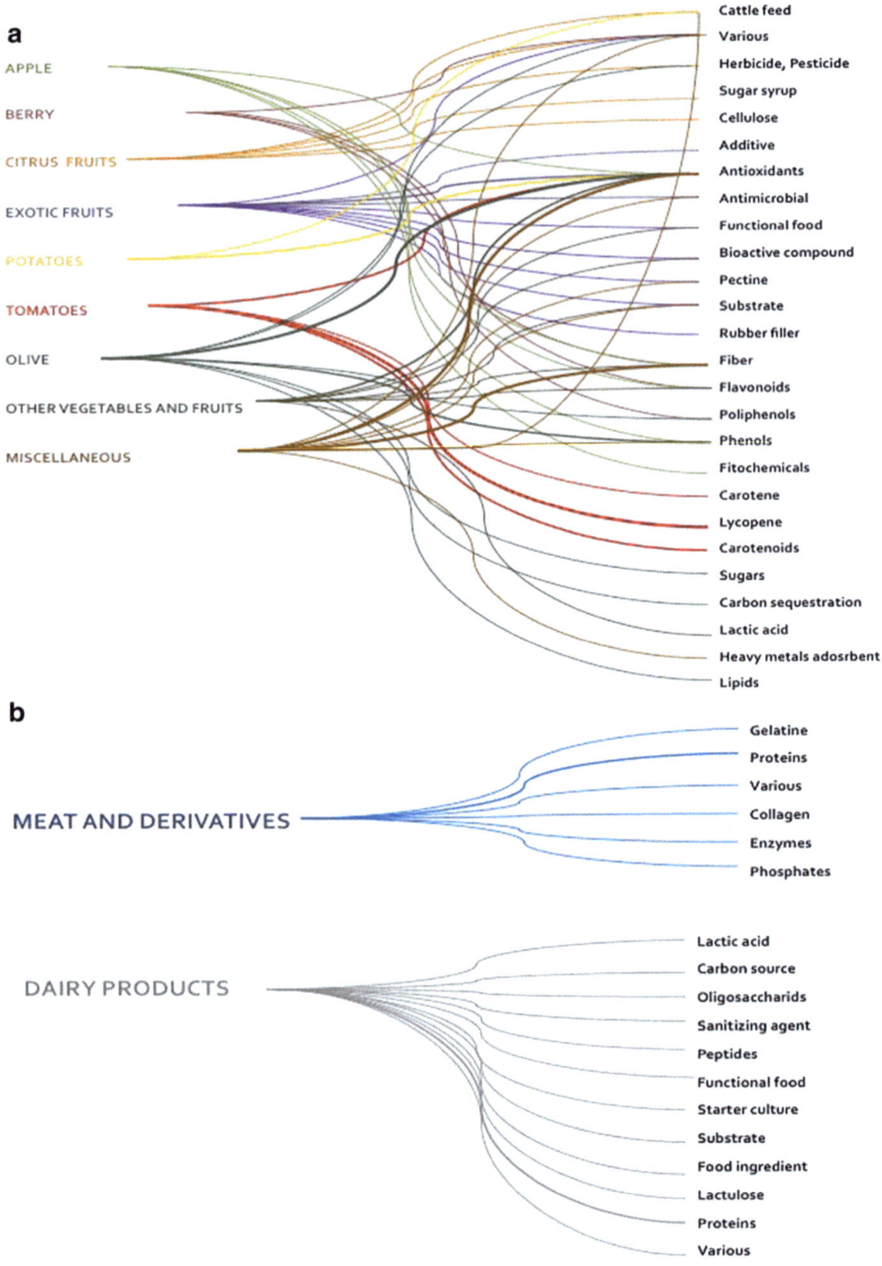

Fig. 3 Food wastage & their potential reuses (*Source* Nadia et al. 2014)

such as corn, peanut, olive, palm, rapeseed, sunflower and soybean.. Although the limits of food vs. energy crunch have stimulated the development of technologies for blending biodiesel from non-food sources such as cotton, nagchampa, neem, jatropha and karanja. Existing life cycle assessment studies prove that utilization of biodiesel as alternative usages of conventional diesel can more protect the environment from GHC emission. The saving percentage by using LCA has been reported but varies depending on the raw material. The EU Renewable Energy Directives (2009/28/EC) in Annex V reported GHC emission saving ranging from 40 to 62% as compare to usage of virgin diesel, while 88% saving are given by the usage of waste vegetable as well as animal oil biodiesel.

6 Life Costing Approaches to Food Waste and Its Valorization

The specific purpose of Life Cycle Costing is to measure the entire cost in economic terms (as suggested to multidimensional effect in earlier described for LCS/LCSA) to the product on its whole life cycle. Throughout the analysis process dissimilar deviations was observed from conventional life cycle costing which is common (De Menna et al. 2016). Besides considering other value chain stages conventional life costing techniques covers only that costs which incurred within the company during analysis. This cost typically includes three major cost components which are initial investment, operating costs and resale or disposal costs. The investment costs includes planning, site acquisition, development of design, buildings' construction, procurements and installation of essential equipment, whereas operating cost includes repair and maintenance, energy and exergy, watering, insurance and taxes. The disposal or resale costs comprises on inspection, bulldozing and other government levies (Kim 2011; De Menna et al. 2018).This Life costing formula is reflected in Fig. 4.

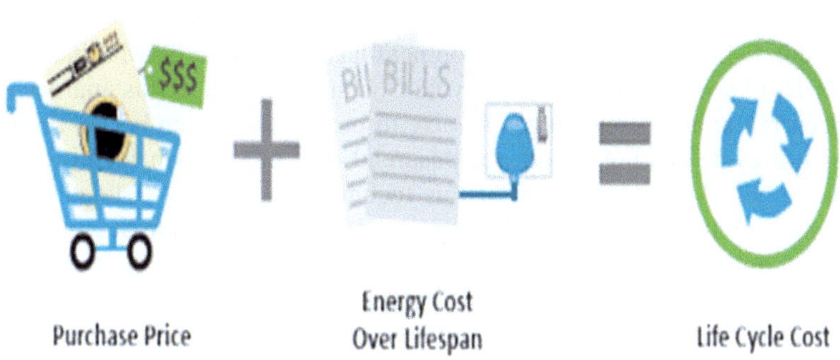

Fig. 4 Life costing formula (*Source* De Menna et al. 2018)

6.1 Environmental Life Cycle Costing (E-LCC)

Recently, Swarr et al. (2011) proposed a most variant technique which is called Environmental Life Cycle Costing (E-LCC) based on life costing cycle and life costing assessment through assigning economic values to the effect elements discovered by coextensive life cycle assessment. As a precedent, Daylan and Ciliz (2016) employed the combination of both Environmental Life Costing and Cost Benefits Analysis to examine the valorization of waste (straw) from wheat crop's waste towards bioethanol demonstrating and found 47% drop in greenhouse gasses pooled with 56% decrease in cost of production admitting a greater risk of eutrophication and photochemical ozone depletion. Conversely, the procedure in which economic values are allocated to environment effects is not permanently forthright and is called into question by several authors (De Menna et al. 2018; San Martin et al. 2020; Guinée 2016; Martinez-Sanchez et al. 2016; Daylan and Ciliz 2016; Cinelli et al. 2013; Swarr et al. 2011; Kim 2011; Guinee et al. 2010; Reich 2005).

A book on Environmental Life-cycle costing (E-LCC) code of practice was published by the society of environmental Toxicology and Chemistry (SETAC) that contains a framework for evaluating decisions with consistent within the boundary of flexible system as an element for sustainable assessment of product. This book based on earlier efforts made by SETAC-Europe Working Group on Life- Cycle Costing summarized within three years span of time (Hunkeler et al. 2008). This code of practice has been developed on a conceptual framework of Life Cycle Sustainability Assessment (LCSA) for the goods which utilized in distinct investigation in every pillar of sustainability, i.e., economy, social equality and environment. The formula for calculating Life Cycle Sustainability Assessment (LCSA) is reflected in previous figures.

6.2 Social Life Cycle Costing (S-LCC)

Beyond this E-LCC an advance development has been made by Martinez-Sanchez et al. (2016) known as societal Life cycle Costing (S-LCC) in this approach entire costs occurred at all society stages involve in project have been borne. Principally, societal life cycle costing is more widespread cost mechanism identified by this approach albeit the methodology of this technique is still not defined very well. Resultantly, innumerable authors have pointed out some complexities while interpreting and implementing the outcomes derived from all kinds of Life cycle costing especially environmental life cycle costing and societal life cycle costing. However, these are best from methodological point of view while deciding how to apply a valorization process in most effective way.

Some authors developed a social impact assessment models for product based on S-LCA techniques. By employing and reviewing these case studies, main factors that affect the product has been established. Earlier in this decade, Maink et al conducted a study by developing S-LCA approach on palm biofuel manufactured in Jambi province at Indonesia. Research results revealed that devoted-active labor relationship played significant role in sustainable development of palm biofuel. The laborers and domestic populations bear a significant social cost for the development made in industrial sector. Similarly Aparcana and Salhofer (2013), in the year 2014 conducted a study by employing S-LCA technique to examine the social effect of garbage assortment and it's recycling on developing countries, three stage valuation index method of garbage assortment as well as it's recycling in the context of developing nations based on S-LCA method and concluded that social effect was estimated by 26 semi quantitative indexes. This approach was also tested as management tool in the study of Gabriella Arcese et al. (2013) on tourism and its trade. The potential social effect has been examined. The questionnaire was developed as per guidelines. The study finding defined the important factors that have social impact on tourist trade.

6.2.1 Prospect and Deficiency

After the issuance of guidelines from UNEP/SETAC in 2009 societal life cycle costing attained a remarkable evolution and development. But this development is still at primary level (Jorgensen 2013), as per proceeding publications, there are some deficiencies of these techniques which are listed below:

a. **Data gathering problems**

Although some statistical year books and other related departments are offering to share the data yet it is not enough. The data gathering is still most serious issue of this technique (Benoit et al. 2010). Presently there is an urgent need of S-LCA database concerning with impact and modeling data that leave scholars facing a lot of difficulties.

b. **Inaccessibility of characterized Models**

The demerit of S-LCA as compared to LCA is unavailability of classified models. As a precedent, the LCA has a characterized models in order to quantitatively combine the environmental effect on human injury eigen value (Guinee et al. 2011), while S-LCA has not this advantage.

c. **Undefinable effect in functional units**

The above elaboration shows that the outcome of S-LCA cannot be computed equally, owing to unavailability of characterized models. Hence, the effect of every operational units cannot be established, that creates hurdle in application of range so the results are inadequate (Hosseinijou et al. 2014).

d. **Software availability problem**

By the help of sufficient data, the sophisticated software can make the analysis possible for social spots as well as universalization of inventory make further simpler and reliable (Lehmann et al. 2013; Benoit et al. 2010).

6.3 Cost–Benefit Analysis

Key aim of cost –benefits analysis which has marginally varies from LCC, by this method company set out and compare end benefits availed from various potential strategies, instead of concentrating on outcomes is most appropriate cost-effective approach for executing a preset strategy. These benefits can be availed through monetary term, environmental or social and they may be realistically adopted the company by applying the CBA (Benis et al. 2018; Jamasb et al. 2010; and Lee et al. 2017). Similar to LCC cost and benefits are normally clustered around preliminary cost set up such as erection of building, procurement of equipment running finance, exergy, labor and end life cost (Tesfaye et al. 2018; Demichelis et al. 2018 and Christoforou et al. 2016). Each of these costs is then matched against the estimated prices of the goods for each projected year of life and afterward reduced to keep in mind the usual relatively high industry preference for an instant rate of return (ROI) offering Net Present Value (NPV) and Benefit–Cost Ratio. (BCR) (Dimou et al, 2016). The bigger it is, generally, the more stable the project is.

7 LCA of the Food Supply Chain

The rising global demand for food has forced intense pressure on the agri-food industry. In order to meet this unprecedented demand, agri-food supply chains are becoming the most important and global as well as characterized by increased exports and imports and the nationwide supply of food products.(WTO 2009; Wognum et al. 2011). The agri-food sector has an active use of resources and its influence on resources and sustainability cannot be denied. This is especially the case in Pakistan; where Pakistan's food sector is the main manufacturing sector transfiguring 70% of agri-products while employing about 56% of the population of SMEs. The food industry and agricultural both are interdependent sectors the agricultural sector ensuring that economical food can be produced in huge quantity by employing scare resources to provide enhanced populace beside extraordinary increased in the prices of raw material. Expectations for increased production and optimum use of scarce resources are growing to satisfy world food demand as population growth is projected to reach 9.60 billion by 2050 (Krishnan et al. 2020). Waste is a major threat to food security, which is still a major issue facing the world, with some 20–30% of the total food being wasted during Post-harvesting supply chain, mainly in developing nations (Krishnan et al. 2020). Once the food is wasted/lost other resources consumed in the manufacturing and distribution process together with the supply chain (FSC) like as water, power, fertilizers, raw materials and pesticides are emaciated too (Geobel et al. 2015). Such widespread use of resources and tremendous waste has a huge negative effect on natural capital, income, humanity, and ecosystems. (Gokarn and Kuthambalayan 2017). These Food Supply chains are also accountable for 31% of greenhouse gas emission as well as Over 50% of eutrophication (Tukker and Jansen 2006). Environmental effect is reflected in Fig. 5

Kirshnan et al (2020) conducted on the mango FSC resulting from the pulp production industry. This study aims to measure the ecological effect of the mango FSC, considering all the resources used, such as planting, manufacturing, packaging, and transporting LCA to mango pulp with its supply chain, in four phases, which are reflected in Fig. 6.

The study concluded by apprising the environment impact of mango pulp on all above four stages that are reflected in tabulated form as below (Table 2).

For this study inventory data was gathered and analyze its effect assessment on environment also estimated (ISO 2006). The effect categories used in this study was earlier recommended in the data system book of International Reference Life Cycle (Wolf 2010).The CML-IA software database was used for impact assessment with its eight categories already established for measurement bases by the Center for Environmental Sciences Leiden University (Guine et al. 2002). Only those categories

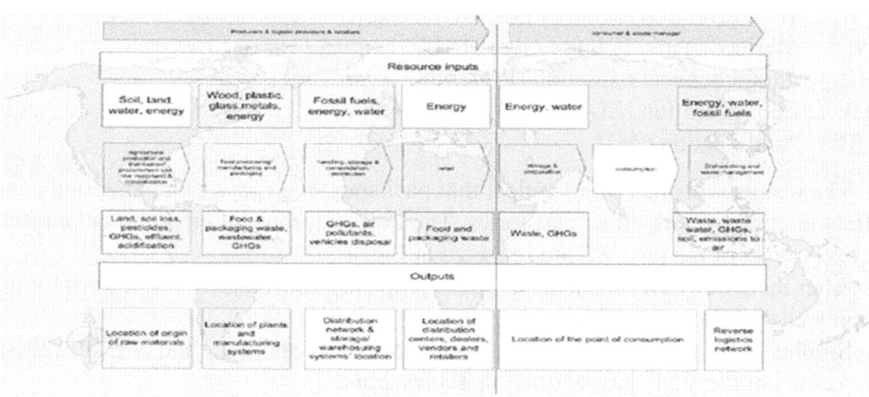

Fig. 5 FSC Environmental impacts (*Source* EEA 2010)

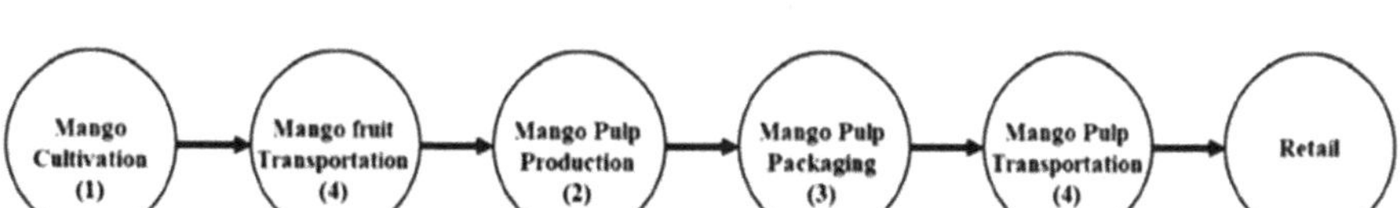

Fig. 6 Mango food supply chain (*Source* Krishnan et al. 2020)

Table 2 Environmental effects corresponding to different phases of mango FSC for 215 kg pulp drum

Effect categories	Unit (Kg)	Total	Effect on cultivation	Effect on pulp production	Effect on pulp packaging	Effect on transportation
GWP100a	CO_2 eq	143.695	42.96	15.820	51.924	32.991
FW	1,4-DB eq	8.378	6.734	0.201	1.1764	0.258
HT	1,4-DB eq	1277.01	26.849	2.672	1239.64	7.851
TE	1,4-DB eq	2.834	1.308	0.024	1.474	0.028
MAE	1,4-DB eq	53,664	7646	11,826	31,153	3037
PO	C_2H_2 eq	0.040	0.001	0.004	0.009	0.018
EU	PO4- eq	0.432	0.345	0.013	0.026	0.048
AC	SO_2 eq	1.209	0.254	0.113	0.346	0.496

Source Krishnan et al. (2020)

were selected which measure the impact of the FSC on natural resources, the natural environment and human health (ISO 2006). These categories are often discussed in the LCA's Food Supply Chain Literature (Manfredi and Vignali 2014; Strazza et al. 2015). These categories are.

I. Global Warming Potential (GWP100a)
II. Fresh Water aquatic Eco toxicity (FW)
III. Human toxicity (HT)
IV. Terrestrial Eco toxicity (TE)
V. Marine aquatic Eco toxicity (MAE)
VI. Photochemical oxidation (PO)
VII. Eutrophication (EU)
VIII. Acidification (AC)

The results plots in Table 1 reflect that packages stage pays more environmental effect in almost entire effect categories. The Potential roots of environmental effect at every level of mango FSC are reflected at Fig. 7.

After their study, Krishnan at el. (2020) proposed a sustainable framework for the supply chain of mango as shown in Fig. 8.

Similar study on tomato purée was performed by Manfredi and Vignali (2014) the results of the study are reported in Tables 2 and 3.

From Table 2 it can be easily concluded that the package stage reflects more environmental effects in almost all of the categories. The LCA application for mango pulp and tomato puree is a precedent that the packaging stage in the FSC is very limited and has a significant environmental impact.

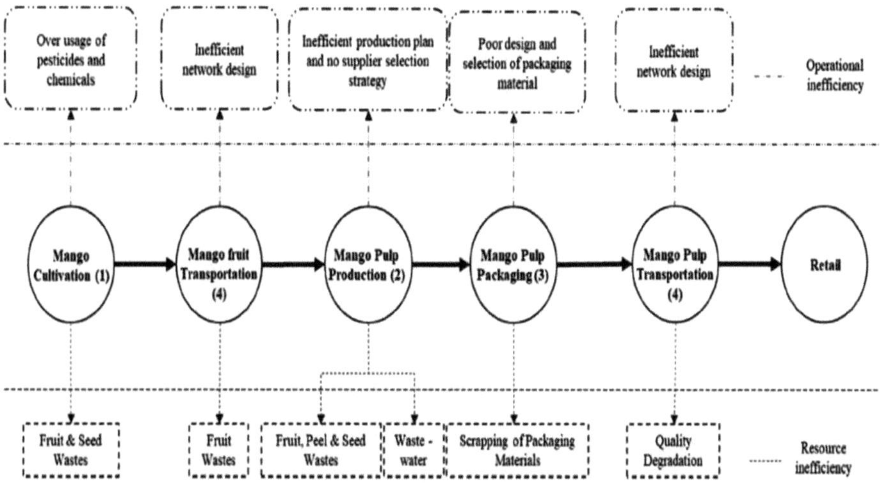

Fig.7 Possible reasons of environmental effect at entire level of mango FSC. *Source* Krishnan et al. (2020)

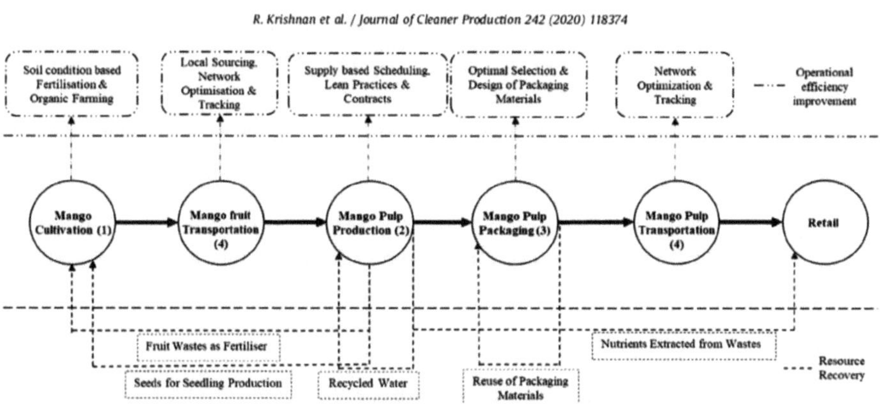

Fig.8 Supply chain of mango (*Source* Kirshnan et al. 2020)

7.1 Current Efforts on LCA

In the developed nations mass food waste is at the later stage of the supply chain at the consumer end. Standard food waste of developed nations such as Europe varies from 95 to 115 kg annually (FAO 2011). In the EU context, about 50% of the food is wasted at the end of the customer or by the households. In addition to the other food-related industries, food services lost 12%, food processing as 19%, full sales, and retail 5%, and production only 11% (Stenmarck et al. 2016). Approximately seventy-five percent of this waste is handled in a wind turbine or open-air amenities

Table 3 Environmental effects corresponding to different phases of Tomato Puree FSC

Effect categories	Unit (Kg)	Total	Effect on processing	Effect on cultivation	Effect on packaging	Effect on transportation
CED	MJ	9.65	1.79	0.94	5.17	1.76
GWP100a	CO_2 eq	0.674	0.108	0.181	0.274	0.111
POP	C_2H_4 eq	0.0001	1.10	7.840	7.760	1.510
ODP	CFC-11 eq	7.120	1.140	1.050	3.160	1.770
AP	SO_2 eq	0.003	0.0002	0.0003	0.002	0.0005
EP	PO_4 eq	0.002	4.290	0.0014	0.0004	0.0001
PM	PM_{10} eq	0.001	5.11	9.80	0.001	0.0002
IR	U_{235} eq	0.0434	0.0011	0.0033	0.029	0.011
WF	Liters	104.90	1.3	103.60	-	-

Source Manfredi and Vignali-2013

(Burns and Hopwood 2017). The anaerobic digestion process is not appropriate for the handling of woody green waste and thus requires mixed planting and in-vessel composting techniques.

This food wastage has significant stress on food sustainability for our system. Multifaceted and creative solutions are constantly being proposed, tested, and implemented by researchers, non-governmental organizations, food industries, and government bodies to address the issue of food loss. Bug-based bioconversions are described as an extravagant solution to minimize food waste/loss. This relatively unique approach can precisely redirect various tons of food lost into valued foodstuff containing animal feed, human food, fertilizer as well as other ancillary manufacturing compounds (Ojha 2020). The above three food waste management techniques are further outlined as under.

7.2 *Anaerobic Digestion (AD) Method*

Anaerobic digestion (AD) is an enticing and widespread technology for improved waste management and energy recovery from food waste. Optimizing the process in-depth understanding of the microbiology associated with this approach is a prerequisite for future operational performance (Jonge et al. 2020). This process is often referred to as biogas processing and biogasification, which is a flexible technology that can be of great benefit to society and industry. As an example a large amount of sludge is created during the treatment of wastewater. It is now becoming expensive to dispose of this waste, and the use of AD will reduce the amount of sludge efficiently. In addition, renewable exergy in form of biogas is extracted through microbial decomposition of the organic component of the sludge, greatly reducing wastewater treatment costs. Anaerobic digestion is devoted to the production of energy. In view

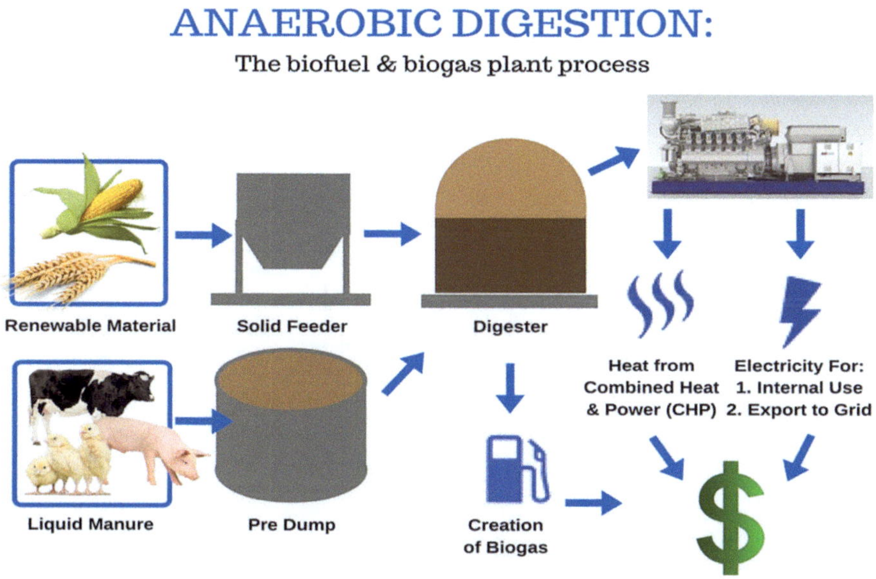

Fig. 9 Anaerobic digestion (*Source* Gulf Coastal Environmental system 2018)

of the advantages of this method, the European governments are now considering applications for DA in the agricultural sector; the purpose behind this application is to reduce the environmental effects of agricultural operations, in addition to increasing the consumption of nutrients used in crop cultivation. Lastly, also the treatment of municipal wastes' organic fraction and organic industrial waste is effectively accomplished through this AD system (Madsen et al. 2010). This is not the intention of this research to recite a lot of interesting responses as well as phenomena occurring at the implementation of the AD process. However, for the clarity of the readers, a clear description of the AD flow chart referred to in a variety of studies is reflected in Fig. 9.

7.3 In-Vessel Composting (IVC) Method

In this process, the food wastage is shredded and placed into the revolving drum composts. The waste is held in these containers for four days. Four days later, this compost is then steam-pasteurized at a temperature of 70 °C in order to comply with United Kingdom regulation (UK government, 2014). This compost is then matured in open-air windows prior to utilize as fertilizer for the agricultural sector. This IVC technique also has the added advantage of displacing the same volume of mineral fertilizers. Biofilters and wet scrubbers are applied in this IVC system to remove odors and pollutants, espically hydrogen sulphide and ammonia. This IVC system

Fig. 10 In-vessel composting (IVC) Method (*Source* Whirlston 2018)

consumed a high voltage of electricity due to an open-air handling system to monitor and preserve the environment for further information, read the Slorach et al. report (2019). The flow chart of the IVC methods referred to in several studies is shown in Fig. 10.

7.4 *Insect Based Bioconversions Method*

According to Global Market Insight, the international demand for edible pests is expected to grow by about 47% between 2020 and 2026 due to rising protein demand, changing nutritional needs, and rapid penetration of edible pests on the global marketplace. Market demand for edible pests is estimated to have gained significant importance over the projected timeline from more than USD 1.5 billion by 2026 (Ahuja and Mamtani 2020). Yet, according to the study by Bakalis et al. 2020, the creation of the emerging FSC in 2020 will entail a balance between the existing world, food supply chains, and other domestic and trends, and will eventually end with glocal strategies. Several companies are leaders in the research and production of protein-based pests for food and feed. Currently, LCA has been used to quantify the environmental effects of pests on feed and food processing processes. Revolutionary research that applied LCA to pests for public consumption was published in 2012 (Oonincx and De Boer 2012) following previous studies on pests for fish feed, animal feed, and organic fertilizers. Lately, Smetana et al. (2019) introduced LCAs for food industry

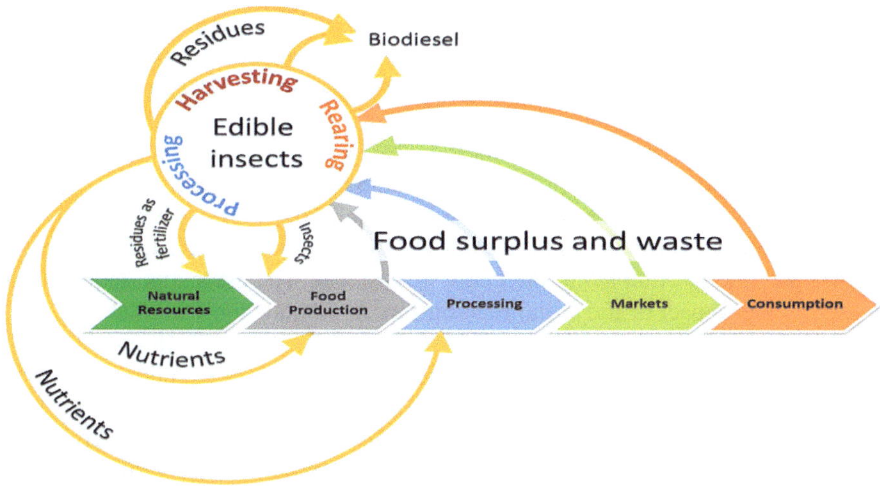

Fig. 11 Insect based bioconversions method (*Source* Whirlston 2018)

side-stream transformation by applying Hermetiaillucens to intermediate food and feed products. In addition to the general ideas of pest processing and manufacturing, new comprehensive data sets were observed, as well as the production of raw materials, treatment, and storage of the hermetiaillucene growth cycle of feedstuffs (from larvae hatching, egg production larvae harvesting, and growing) The method of ultimate yields incorporates multiple products such as protein concentrate, fresh insect purity, organic fertilizer, and pest fat. This method was considered to be the most effective experimental plant with insight into potential and up-scaling scenarios. The process flow chart of this system can be found in the Fig. 11.

8 Conclusion

This chapter introduces the evaluation, emergence, origin, classification framework work as well as application of the Life Cycle Assessment (LCA) to sustainable food production in detail and covers the entire food supply chain analysis of food production and its impact on environmental degradation.

The LCA is an environmental assessment procedure that aims to assess the impact of food waste from generation to generation. This study also analyses and learns about the important ingredients in food waste that can produce valuable products such as bio-fuel, biogas and other valuable products. This chapter also provides a case study of examples of Coca-Cola bottles that have been evaluated in the industry and its outcome has also been given that LCA has attained maturity and delivers an imperative role in ecological assessment, particularly in entire food supply chain.

The economic impact has been concentrated and its application is appropriate in the field of economics, it minimizes the acquisition costs. Stakeholders can make use of the benefits of the situation, taking into account the long-term benefits. By introducing LCA to the entire supply chain system, the optimal utilization of resources, sustainable benefits, other by-products and can save the environment from degradation as well as other valuable products through the process of food waste recycling. The assessment of the life cycle and the cost of the life cycle of both approaches are not adequate to achieve the maximum sustainability goal. The S-LCA originates for the purpose of completing the entire assessment. The estimation target of the S-LCA is the impact of the social relationship, in conjunction with material activities, as well as other activities carried out by stakeholders, which contributes to further development of the LCSA. Food loss and waste is a significant barrier to achieving sustainability goals and saving the population of developing nations from starvation. Multifaceted and creative ideas are continuously being proposed, tested, and implemented by researchers, non-governmental organizations, food companies, and government bodies to resolve the problem of food loss. Three LCA processing techniques, e.g., anaerobic digestion (AD), In-vessel composting (IVC), and insect-based bioconversions, are deeply discussed in this chapter to explain the reader flow diagram of each form. At that same time, the bioconversion technique based on insects is most creative and suitable to meet the increasing need for proteins.

References

Ahuja K, Mamtani K, (2020) Edible insects market size to exceed $1.5bn by 2026. Global Market Insights Inc

Aparcana S, Salhofer S (2013) Application of a methodology for the social life cycle assessment of recycling systems in low income countries: three Peruvian case studies. Int J Life Cycle Assess 18(5):1116–1128

Arcese G, Lucchetti MC, Merli R (2013) Social life cycle assessment as a management tool: methodology for application in tourism. Sustainability 5(8):3275–3287

Ayala-Zavala J, Rosas-Domínguez C, Vega-Vega V, González-Aguilar G (2010) Antioxidant enrichment and antimicrobial protection of fresh-cut fruits using their own byproducts: looking for integral exploitation. J Food Sci 75(8):R175–R181

Bakalis S, Valdramidis VP, Argyropoulos D, Ahrne L et al (2020) How COVID-19 changed our food systems and food security paradigms. Curr Res Food Sci 3:166–172. https://doi.org/10.1016/j.crfs.2020.05.003

Banerjee A, Duflo E, Glennerster R, Kinnan C (2015) The miracle of microfinance? Evidence from a randomized evaluation. Am Econ J Appl Econ 7(1):22–53

Bampidis V, Robinson P (2006) Citrus by-products as ruminant feeds: a review. Animal Feed Sci Technol 128(3–4):175–217

Benis K, Turan I, Reinhart C, Ferrão P (2018) Putting rooftops to use–a cost-benefit analysis of food production vs Energy Generation under Mediterranean Climates. Cities 78:166–179

Benoît C, Norris GA, Valdivia S, Ciroth A, Moberg A, Bos U, Prakash S, Ugaya C, Beck T (2010) The guidelines for social life cycle assessment of products: just in time! Int J Life Cycle Assess 15(2):156–163

Broad R, Cavanagh J (1993) Plundering paradise: the struggle for the environment in the Philippines. University of California Press

Burns C, Hopwood L (2017) Organics recycling industry status report 2015. WRAP. http://www.wrap.org.uk/sites/files/wrap/asori%202015.pdf

Camp LA, Hofmann SL (1993) Purification and properties of a palmitoyl-protein thioesterase that cleaves palmitate from HRas. J Biol Chem 268(30):22566–22574

Cerda A, Artola A, Font X, Barrena R, Gea T, Sánchez A (2018) Composting of food wastes: status and challenges. Bioresour Technol 248:57–67

Ceren R, Doshi P, Banerjee B (2016) Reinforcement learning in partially observable multiagent settings: monte carlo exploring policies with PAC bounds. 530–538

Cherubini F, Strømman AH (2011) Life cycle assessment of bioenergy systems: state of the art and future challenges. Bioresour Technol 102:437–451

Christoforou E, Kylili A, Fokaides PA (2016) Technical and economical evaluation of olive mills solid waste pellets. Renew Energy 96:33–41

Cinelli M, Coles SR, Jørgensen A, Zamagni A, Fernando C, Kirwan K (2013) Workshop on life cycle sustainability assessment: the state of the art and research needs—November 26, 2012, Copenhagen, Denmark. Int J Life Cycle Assess 18(7):1421–1424

Daylan B, Ciliz N (2016) Life cycle assessment and environmental life cycle costing analysis of lignocellulosic bioethanol as an alternative transportation fuel. Renew Energy 89:578–587

De Menna F, Malagnino RA, Vittuari M, Molari G, Seddaiu G, Deligios PA, Solinas S, Ledda L (2016) Potential biogas production from artichoke byproducts in Sardinia, Italy. Energies 9(2):92

De Menna F, Dietershagen J, Loubiere M, Vittuari M (2018) Life cycle costing of food waste: a review of methodological approaches. Waste Manage 73:1–13

Demichelis F, Fiore S, Pleissner D, Venus J (2018) Technical and economic assessment of food waste valorization through a biorefinery chain. Renew Sustain Energy Rev 94:38–48

Dimou C, Vlysidis A, Kopsahelis N, Papanikolaou S, Koutinas AA, Kookos IK (2016) Techno-economic evaluation of wine lees refining for the production of value-added products. Biochem Eng J 116:157–165

Donald PF, Evans AD (2006) Habitat connectivity and matrix restoration: the wider implications of agri-environment schemes. J Appl Ecol 43(2):209–218

Dobbs C, Escobedo FJ, Zipperer WC (2011) A framework for developing urban forest ecosystem services and goods indicators. Landscape and urban planning 99(3–4):196–206

Ebner J, Babbitt C, Winer M, Hilton B, Williamson A (2014) Life cycle greenhouse gas (GHG) impacts of a novel process for converting food waste to ethanol and co-products. Appl Energy 130:86–93

Fao F (2012) Agriculture organization of the United Nations. 2012. FAO Statistical Yearbook

Field T (1992) Wandering souls: disappearing forests. American Forests (USA)

Glavič P, Lukman R (2007) Review of sustainability terms and their definitions. J Clean Prod 15(18):1875–1885

Gokarn S, Kuthambalayan TS (2017) Analysis of challenges inhibiting the reduction of waste in food supply chain. J Clean Prod 168:595–604

Guine J, Gorrée M, Heijungs R, Huppes G, Kleijn R, Udo de Haes H, Van der Voet E, Wrisberg M (2002) Life cycle assessment. An operational guide to ISO standards. Vols, 1–3

Guinee JB, Heijungs R, Huppes G, Zamagni A, Masoni P, Buonamici R, Ekvall T, Rydberg T (2011) Life cycle assessment: past, present, and future

Guinée J (2016) Life cycle sustainability assessment: what is it and what are its challenges? In taking stock of industrial ecology. Springer, Cham, pp 45–68

Gustafsson J, Cederberg C, Sonesson U, Emanuelsson A (2013) The methodology of the FAO study: global food losses and food waste-extent, causes and prevention"-FAO, 2011

Homer-Dixon TF (1991) On the threshold: environmental changes as causes of acute conflict. Int Secur 16(2):76–116

Hosseinijou SA, Mansour S, Shirazi MA (2014) Social life cycle assessment for material selection: a case study of building materials. Int J Life Cycle Assess 19(3):620–645

Hunkeler D, Lichtenvort K, Rebitzer G (2008) Environmental life cycle costing. Crocpress

Hunt RG, Franklin WE, Hunt RG (1996) LCA—how it came about. Int J Life Cycle Assess 1:4–7

International Energy Agency IEA, (2014) Key World Energy Statistics. Paris, p 6
IMF Annual Report 2015 (n.d.) Retrieved August 6, 2021 from https://www.imf.org/external/pubs/ft/ar/2015/eng/index.htm
Jackson J (2009) Neo-liberal or third way? What planners from glasgow, Melbourne and toronto say. Urban Policy and Res 27(4):397–417
Jacobson JL (1989) Environmental refugees. Futurist 23:54
Jacobson M (2009) Review of solutions to global warming, air pollution, and energy security. Energy Environ Sci 2(2):148–173
Jamasb T, Nepal R, Kiamil H (2010) Waste to energy in the UK: policy and institutional issues. Proc Inst Civil Eng Energy 163(2):79–86
Jonge N, Davidsson A, Jansen Cour J, Nielsen LJ (2020) Characterization of microbial communities for improved management of anaerobic digestion of food waste. Waste Manage 117:124–135
Kim DH (2011) A review of desalting process techniques and economic analysis of the recovery of salts from retentive. Desalination 270(1–3):1–8
Kiran EU, Trzcinski AP, Ng WJ, Liu Y (2014) Bioconversion of food waste to energy: a review. Fuel 134:389–399
Krishnan R, Agarwal R, Bajada C, Arshinder K (2020) Redesigning a food supply chain for environmental sustainability–an analysis of resource use and recovery. J Clean Prod 242:118374
Lee KH, Oh J, Chu KH, Kwon SH, Yoo SS (2017) Comparison and evaluation of large-scale and on-site recycling systems for food waste via life cycle cost analysis. Sustainability 9(12):2186
Lin CSK, Pfaltzgraff LH, Davila L, Mubofu E, Abderrahim S (2013) Food waste as a valuable resource for the production of chemicals, materials and fuels. Current situation and global perspective. Energy Environ Sci 6(2):426–464
Madsen M, Holm JB, Esbensen KH (2010) Monitoring of anaerobic digestion processes: a review perspective. Renew Sustain Energy Rev 15:3141–3155
Manfredi M, Vignali G (2014) Life cycle assessment of a packaged tomato puree: a comparison of environmental impacts produced by different life cycle phases. J Cleaner Production 73:275–284
Martinez-Sanchez V, Tonini D, Møller F, Astrup TF (2016) Life-cycle costing of food waste management in Denmark: Importance of indirect effects. Environ Sci Technol 50(8):4513–4523
Meadows D, Randers J, Meadows D (2004) Limits to growth: the 30-year update. Chelsea Green Publishing
Mena C, Adenso-Diaz B, Yurt O (2011) The causes of food waste in the supplier–retailer interface: evidences from the UK and Spain. Resour Conserv Recycl 55(6):648–658
Morero B, Rodriguez MB, Campanella EA (2015) Environmental impact assessment as a complement of life cycle assessment. Case study: Upgrading of biogas. Bioresour Technol 190:402–407
Myers JR (1997) Injuries among farm workers in the United States, 1993. US Department of Health and Human Services, Public Health Service, Centers
O'Brien C (1999) Sustainable production–a new paradigm for a new millennium. Int J Prod Econ 60:1–7
Oonincx DG, De Boer IJ (2012) Environmental impact of the production of mealworms as a protein source for humans–a life cycle assessment. PLoS One 7(12)
Ojha A (2020) Nanomaterials for removal of waterborne pathogens: opportunities and challenges. Waterborne Pathogens 385–432
Paull J (2014) Lord Northbourne, the man who invented organic farming, a biography. J Org Syst 9(1):31–53
Prahalad CK, Ramaswamy V (2004) Co-creation experiences: the next practice in value creation. J Interactive Marketing 18(3):5–14
Reich MC (2005) Economic assessment of municipal waste management systems—case studies using a combination of life cycle assessment (LCA) and life cycle costing (LCC). J Clean Prod 13(3):253–263
Roy P, Nei D, Orikasa T, Xu Q, Okadome H, Nakamura N, Shiina T (2009). A review of life cycle assessment (LCA) on some food products. J Food Eng 90(1):1–10

Sachs JD (2015) The age of sustainable development. Columbia University Press
Salemdeeb R, Ermgassen ZU, Kim EKHJ, Balmford MH, Al-Tabbaa A (2017) Environmental and health impacts of using food waste as animal feed: a comparativeanalysis of food waste management options. J Clean Prod 140:871–880
San Martin D, Orive M, Martínez E, Iñarra B, Ramos S, González N, de Salas AG, Vázquez LA, Zufía J (2020) Multi-criteria assessment of the viability of valorising vegetable by-products from the distribution as secondary raw material for animal feed. Environ Sci Pollut Res 1–15
Shalini R, Gupta D (2010) Utilization of pomace from apple processing industries: a review. J Food Sci Technol 47(4):365–371
Sen S, Bhattacharya CB (2001) Does doing good always lead to doing better? Consumer reactions to corporate social responsibility. J Mark Res 38(2):225–243
Slorach PC, Jeswani HK, Cuéllar-Franca R, Azapagic A (2019a) Environmental and economic implications of recovering resources from food waste in a circular economy. Sci Total Environ 693:133516
Smetana S, Schmitt E, Mathys A (2019) Sustainable use of Hermetia illucens insect biomass for feed and food: attributional and consequential life cycle assessment. Resour Conserv Recycl 144:285–296
Stenmarck AS, Jensen C, Quested T, Moates G, Buksti M, Cseh BZ, Juul S, Parry A, Politano A, Redlingshofer B (2016) Estimates of European food waste Levels. FUSIONS - European Commission
Strazza C, Magrassi F, Gallo M, Del Borghi A (2015) Life Cycle Assessment from food to food: a case study of circular economy from cruise ships to aquaculture. Sustain Prod Consumpt 2:40–51
Suhrke A (1994) Environmental degradation and population flows. J Int Affairs 473–496
Swarr TE, Hunkeler D, Klöpffer W, Pesonen H.-L, Ciroth A, Brent AC, Pagan R (2011) Environmental life-cycle costing: a code of practice
Tesfaye T, Johakimu JK, Chavan R, Sithole B, Ramjugernath D (2018) Valorisation of mango seed via extraction of starch: preliminary techno-economic analysis. Clean Technol Environ Policy 20(1):81–94
Tseng M-L, Divinagracia L, Divinagracia R (2009) Evaluating firm's sustainable production indicators in uncertainty. Comput Ind Eng 57(4):1393–1403
Tukker A, Jansen B (2006) Environmental impacts of products: a detailed review of studies. J Ind Ecol 10(3):159–182
Wognum PN, Bremmers H, Trienekens JH, Van Der Vorst JG, Bloemhof JM (2011) Systems for sustainability and transparency of food supply chains–current status and challenges. Adv Eng Inform 25(1):65–76
Wolf M (2010) Forging forward with 10 burning questions on FGF23 in kidney disease. J Am Soc Nephrology 21(9):1427–1435
Xue L, Liu G, Parfitt J, Liu X, VanHerpen E, Stenmarck Å, O'Connor C, Östergren K, Cheng S (2017) Missingfood,missingdata?Acriticalreviewofglobalfoodlosses

Sustainable Production Practices and Future Trends

Adeel Shah, Zhang Yu, and Pablo Ponce

Abstract Sustainable production practices are of vital importance to achieving sustainability. The future trend practices base on technological advancement in communication technology and materials. Companies across the globe are targeting to reduce their carbon footprint to achieve carbon neutrality. The sustainable production practices of the future will require collective wisdom and knowledge management. Sustainability would be futile if performed acutely by a company in a specific location. Instead, it is only possible if the complete supply chain's stakeholder synergies for the cause. The triple bottom line is a useful concept that provides businesses with measurable indicators to monitor their activities. Sustainable practices future trends are evident to provide companies with higher efficiencies in production while concurrently reducing any negative impact on society and the environment. The responsible companies of the future will intelligently tradeoff between manufacturing resources, robotics, and human resource. The practices brought forward in the chapter as future trends are to achieve United Nations 17 sustainable development goals for the year 2030.

Keywords Sustainable production · Sustainability · Industrial ecology · Future trends

A. Shah (✉)
Collage of Business Management, Institute of Business Management, Karachi, Pakistan
e-mail: adeel.shah@iobm.edu.pk

Z. Yu
School of Economics and Management, Chang'an University, Xi'an, China

P. Ponce
Carrera de Economía, Universidad Nacional de Loja, Loja, Ecuador

C. J. Chiappetta Jabbour and S. A. R. Khan (eds.), *Sustainable Production and Consumption Systems*, Industrial Ecology,
https://doi.org/10.1007/978-981-16-4760-4_6

1 Introduction

The idea of sustainability has emerged since the 1700s in Germany. The crown for being 'father of the subject of sustainable development goes to Hans Carl von Carlowitz (Lees 2012). Hans Carl used the sustainable development concept to grow timber by sowing seed to reduce shortages due to increase consumption. However, many researchers since the 1900s have tried to pin down the definition of sustainable development. The definition widely quoted is by Brundtland report 1987 (World Commission on Environment and Development and Brundtland 1987; World Nuclear Org 2020). The definition clearly defines the goal and aims of required sustainability in every walk of life and across the globe. The report stresses manufacturing and consuming responsibly for present human needs while not compromising future generations' needs. Sustainable development in all activities will ensure all human beings are responsible and live up to the Brundtland report's definition. The whole sustainability concept stands on three pillars: social, environmental, and economical; our actions should ensure that no negative unto them (see Fig. 1).

The sustainability progress starts with the Brundtland report and completes United Nations seventeen Sustainable Development Goals 2030 (U.N. 17 S.D.G.s) (United Nations 2015), accepted by all member countries. The global event led to make manufacturing and service industries responsible and vigilant to reduce any negative impact of their business actions on the three pillars of sustainability. The three sustainability pillars further evolved to fit in business models (see Fig. 2), producing tangible and intangible products. The idea is to optimized the manufacturing process to create sustainable outcomes with less harm to the three pillars discussed earlier in the chapter. Society entails all the company employees, consumers, and the people living in the organization's situated location.

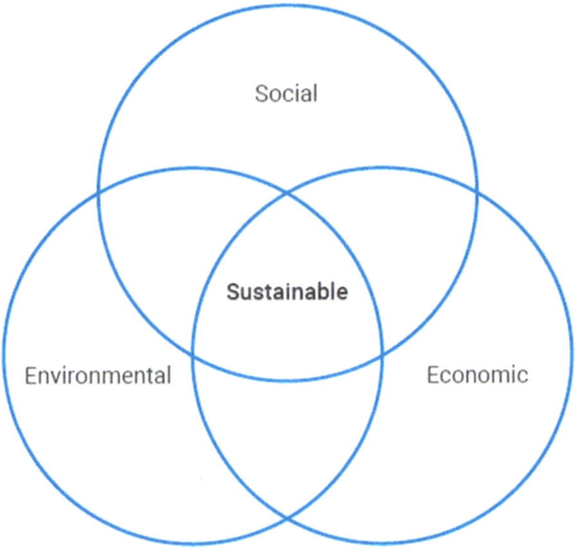

Fig. 1 The three pillars of sustainability (World Nuclear Org 2020)

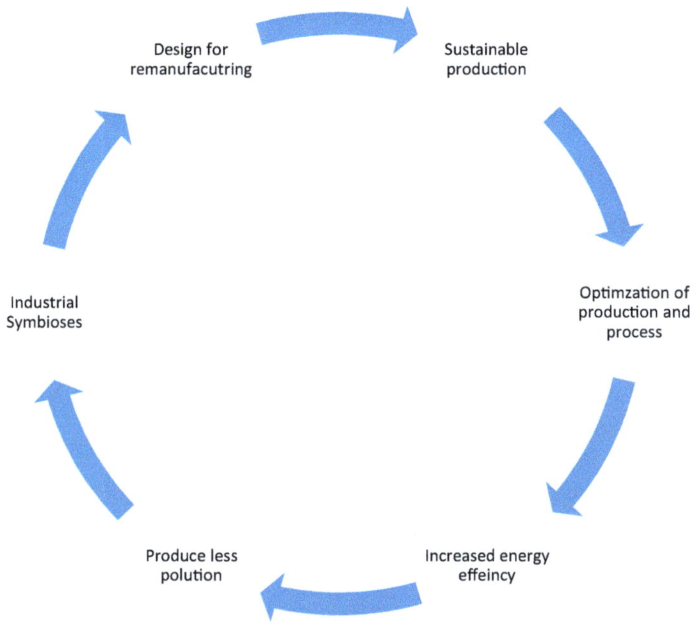

Fig. 2 Sustainable production process (Sustainability Guide 2018)

Interestingly, business organizations worldwide incorporate sustainability in their operations management framework for short and long-term operations (Helen et al. 2014). The organization's sensitivity to sustainability has increased due to the propagation from United Nations and societal awareness and demands. The voices of concerns are from the organization's internal and external stakeholders (employees and customers).

The organizations' sustainability efforts measures through the use of the triple bottom line concept (TBL). The triple bottom line refers to the Planet (environmental), Social (people), and Profit (economic) aspects. The organizations are vigilant and use the lens of TBL to monitor their operations and operational linkages of all stakeholders in the supply chain network. Organizations being vigilant on their supply chains tend to cover all product stages from raw material from the supplier to the final product consumed and reuse the finished product by closed-loop supply chain management strategy (Shi et al. 2021). Sustainability in operations management has forced organizations to rethink product design to be more eco-friendly, sustainable compliance, lean-sensitive, and green supply chain. The organizational culture has evolved in the past 20 years: as in the 1990s, the focus was to align resources to become more productive and reduce resource consumption. However, the current pace would take more than one earth resource to meet the demand (Weizsäcker et al. 1998). The trends have now changed to more renewable resources to strive for having greener business activities and remain competitive to earn profits. Environmental performance is a significant focus by organizations and government environmental regulatory bodies, resulting in new frameworks and good practices to reduce waste in process and carbon emissions. The trends in sustainable production is now more focus on the triple bottom line intertwined with business operations and its management.

2 Future Trends in Sustainable Production

The future trends in sustainable production will encompass all three Ps. The organizational are learning fast and trying to improve their methods to conduct business. Parallel to these business efforts, researchers and engineers are also thinking, improving, and proposing ideas to meet the SDGs goals of United Nations. Given the latest developments, below are some trends in sustainability.

3 Efficiency in Resources Consumption

Resource consumption during the past century has increased dramatically. The resource consumptions include mineral resources, fossil fuels, and agriculture products. The reason behind this steep increase in consumption is due to the rising population and income. The population trends are most worrisome, especially in developing and under-developed countries. According to United Nations (2020) estimates, in the year 2030, the population will grow to 8.5 billion.

Further, in 2050, it will grow to 9.7 billion, and in the year 2100, it will grow to staggering numbers of 11 billion. The numbers reflect that organizations strive for sustainable production; the consumers have to adopt sustainable consumption patterns. The gaps till to-date of developed, developing, and underdeveloped countries are wide. The rice is getting richer while the poor are getting poorer. The rich people have increased their consumption which poor are remaining deprived which is another significant concern.

The wealthy of society has to adopt sustainable consumption patterns. The whole equation of sustainable production depends on the ferocious demand arisen due to non-sustenance consumption. The adaptation has to start from the wealthy of the society, and it will have a trickle-down effect. The developed countries have to take the lead in consuming the products and become an example for other countries and nations. The living standard of the poor will rise due to available resources. The trend for responsible consumption is now becoming more visible; the increasing entropy can reduce through it. To back up the cultural change, the business enterprises should use production systems that eliminate any wastages during the process and ensure their global supplier implement the same by giving technical support and incentives, thus starting the culture of sustainable consumption and production globally.

4 Trends Sustainable Manufacturing

The trend to revolutionize sustainable production practices focus on optimizing manufacturing. The optimization tends to reduce the negative impact of the environment's process by conserving energy and resources. Sustainable manufacturing also engulfs employees and community safety from the adverse effects of the process. The optimization of manufacturing will result in better production performance. It entails re-engineering the whole manufacturing process with the internet of things, automation, and closed looped systems. The revamp will reduce cost while increasing production efficiency. Another important goal of sustainable manufacturing as a trending practice is to reduce energy consumption during the process. The scientist and engineers are experimenting with materials to minimize the inertias and weight of the machine component during production; thus, it results in lower energy consumption and reduction in heat production. With renewable resources like carbon fibers and dissolvable plastics, companies' machine components will have a longer life, and after practical use, the parts recycle to use again.

The same part can use as a component in other machines. The sustainable production practices trends are more focused on environmental impact due to waste and pollution emission. Here a discussion on Asia is essential as the continent is a significant producer. India, Pakistan, and Bangladesh are focusing on waste reduction. Simultaneously, China, Japan, Singapore, Malaysia, and few others focus on manufacturing re-engineering for a greener output targeting the triple bottom line in true spirit. The earlier discussed three manufacturing giants are fast to catch up the pace with their other counterparts. Western countries have passed the threshold and are more focused on incorporating technology to become lean. The lean is envelops to consume all resources efficiently with zero wastage as in an ideal situation, and it entails energy, water, raw materials, labor, and finances. Water in the coming years will remain a significant concern in the production process. Textiles are one of the industries where pure water is part of the process and used in abundance. Chemists are trying to innovate the textile process to reduce water usage during production. Freshwater sources are scarce, and the shortage is increasing as the human population grows every passing day.

Additive manufacturing will also become a permanent part of the sustainable production process of the future. The enterprises will use nano raw material for additive manufacturing to reduce production delays and stoppages due to their faulty parts. Additive manufacturing, commonly known as 3D printing, is a technology used on a small commercial scale for product development, product prototype, and finished product. To make it a permanent feature, raw-materials providing good strength required to withstand the fatigues during high-speed manufacturing. The practice will enable to use of recycled plastics and other materials by the organization. Further, it will also reduce transportation impacts on the environment as organizations become self-sustaining for machine parts. Significant savings during product development will become a reality after additive manufacturing evolves to exhibit and mimic the same functional use as the actual production process on the intended raw material.

Lastly, for manufacturing sustainability in production practices, operational safety will also become an essential part of the list. Operational safety encompasses the hazards faced by employees during manufacturing. Many manufacturing processes are causing health issues like respiratory illness, phycological, heart issues, skin problems, and life-threatening in case of fire and explosion. The risk of operational magnitude is different in various industries. It is very high in metal and coal mining, high in textiles, medium in the beauty industry, and low in the automobile industry, depending on its sustainable practices and public awareness. Sustainability to increase operational safety efforts in developed countries is a phenomenon and a bleak situation in developing countries. The manufacturing enterprises in developing and developed countries will rethink to learn from developed countries to safeguard its employee as an essential resource by increasing their operational safety standards in the future.

5 Industrial Symbioses

The other prominent trend that is gaining momentum is the industrial symbioses. The industrial symbioses is an idea to mimic the natural organic cycle in the whole network. The concept revolves around the natural way of degenerating products to release energy and use for other purposes or barnacles and wales. The relationship between barnacles and wales is fascinating as barnacles benefits by attaching to the wales skin without harming it. In industrial settings, the waste from manufacturing a product or faulty component uses as a resource for another process or product. Like in nature, researchers are trying to implement symbiosis in industrial settings, which translates as a collective benefit rather than individual benefit alone. The practical use of the concept is present since the 1970s in Kalundbrog, Denmark. The symbiosis in the Danish city by sharing water sources by four manufacturing organizations. Although from different sectors, a power plant, oil refinery, pharma, and gypsum board facility, the organizations used the same water resources in harmony with depriving each other of the water resource and using byproducts from each different manufacturing process. The Danish companies' impact of symbiosis realized a positive impact on the environment low energy consumption, low carbon emission, minimal wastages, and saving precious water recourse in the locality every year.

As discussed earlier in the sustainable manufacturing, the scientist and researchers are trying to invent a material that reuses or incarnate to produce the same value and utility or provide value elsewhere. Industry 4.0 is a big step in the direction; it gives the framework for new production setups. The concept to design for remanufacturing trends links and complement the idea of industrial symbioses. The products will develop in a manner that their useful components will directly reuse elsewhere. Currently, the concept is known as component standardization which mitigates any excessive or diminishes demand. For instance, use of standard power supply in all

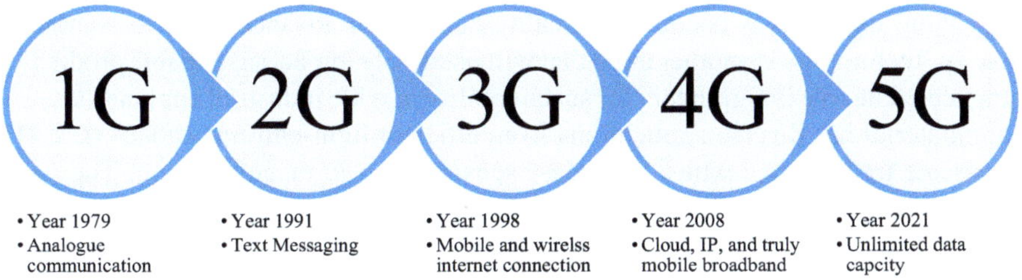

Fig. 3 Evolution of information and communication technologies (Brosowsky 2019)

desktop models sold by companies. However, designing for remanufacturing will broaden the component's utility for a more extended period, thus saving virgin resources in nature.

6 Information and Communication Technologies

The most significant advancement in the twenty-first century has been in information and communication technologies, prominently known as I.C.T. (information and communication technologies). Currently, 5G (Fifth Generation) is the I.C.T. technology which considers a disruptive communication technology 500% faster than our current technology of 4G (Fourth Generation) (see Fig. 3 for I.C.T. evolution). The 5G will enable IoT in sustainable production by allowing devices and machines to communicate faster, thus becoming more reliable for feedback to manage and optimize production. The 5G is a cloud-based technology that ensures better connectivity for machines than its predecessor. Experts claim the technology is truly a neutral carbon network as it provides better-connected devices to mitigate any data losses (Pickup 2020). The I.C.T. currently has a large carbon footprint which will reduce by a factor of 5. The 5G technology will help to conserve energy as many devices will shut when not in use. The environmentalist is optimistic about the technology. It will provide a large amount of data to precisely predict the decision made by organizations and governments to safeguard resources and promote sustainability in the longer run. 5G coupled with IoT will revolutionize sustainable productions to produce more greener products to positively impact the three P's (planet, profit, and people).

7 Nanotechnology

The technology involves manipulating raw material physical and chemical composition on the atomic and molecular size. For better understanding, the scale is one and 100 nm that could translate to a size of marble compared to the earth's size.

Improving materials is becoming a reality because of advancements in Nanotechnology. Technology is another trend that will positively impact sustainable production practices. The role is significant for sustainability in wide industrial applications. The nanomaterial has diverse applications from efficient light-emitting diodes (L.E.D.), nanoscale transistors, textiles, and glass sensors used in manufacturing. The nanomaterial will substitute conventional material to give cleaner and leaner production systems to save energy and raw materials. The advantages will have a broader impact on sustainable production practices, as nanomaterials will enable humans to produce lightweight components in transportation and machines. The component light will make the system fuel and energy-efficient during manufacturing. The technology will improve operational safety during production significantly. The nanomaterial enables the production of safety gears that are light in weight to increase labor efficiency while protecting them from harm's way. The safety gears are possible by dispersing nanosized carbon particles into the original fibers of the garment. The dispersion creates a film of nanofibers on the surface of the thread. Due to the coating of carbon nanofibers, garments are now liquid repellent, stain-resistant and antimicrobial. The other practical use of the technology, which is evident, reduces friction between moving parts of the machines. The technology has unlocked important use as lubricants to replace oils as lubrication and impact the environment by diminishing fossil fuels. The science behind the nanoparticles uses lubrication as it will work as micro bearings or rolling between machine parts to reduce friction.

8 Facility Design

The designing of facilities and infrastructure supplements sustainable production practices and will become a permanent feature for business organizations. Business around the world is designing offices, factories, warehouses and other facilities to increase sustainability. The organizations are sourcing indigenous materials for construction, thus limiting transportation costs and fuel and becoming cost-efficient. Besides construct material, the facilities design to reuse natural resources like water and air by filtration to increase sustainability, wastewater from washrooms, and manufacturing to reuse elsewhere in the process. The practice will reduce freshwater consumption, which can be used to better drinking the community. Architects are becoming creative and bringing out the box ideas to collect rainwater for better use. Lastly, the design reduces electricity consumption by using more daylight for less reliance on electricity-based lighting sources. The concept of natural lighting is now a widespread practice in offices, schools, and manufacturing facilities in developed countries. However, much push is needed to develop and developing countries where electricity is mostly produced from fossil fuel burning. Unfortunately, sunlight free from light is wasted when now ever have material which reduces the heat and filters to pass only light for our use. The natural light is free, clean, and provides a pleasant working environment for employees to improve their efficiency while decreasing operating expenses and higher overall return on facilities' investment.

9 Digital Twins

Digital twin technology in manufacturing is increasing rapidly and will remain a trend in the future. The trends in product development to achieve sustainable production give a positive outlook. In the field of product development, 'Digital Twins' will have a significant impact. The concept is to create a digital copy of the product or component, thus giving it the name 'Digital Twins Modeling.' The model will allow users to fully understand the production workflow and fine-tune the whole process to make it leaner. The forecast maintenance of machinery and other physical assets will be more precise. The engineers paired the two technology to optimize production flows and forecast unexpected shutdowns to avoid any wastage.

All physical products can replicate digitally with the help of digital twin technology. Digital replication is possible from microchips to commercial planes. The technology is used in the automotive industry to given insight into car performance experimenting on different terrains, weather, and temperatures digitally. The other fair use of the technology is for predicting the utility and life span of a vital machine part. Digital twin technology can provide better and accurate estimates to avoid significant repairs unnecessary stoppages. Researchers are using virtual reality (V.R.), augmented reality (A.R.), and donning wearable technology for engineers to see conceal components or parts that would waste hours analyzing for problems or defects. The best example to bring forward is the maintenance team arriving at the site to check to conceal the system (cemented in the wall) with A.R., and the digital twin of the system will replicate and artificially mimic the real-time data to bring forward the current condition of the system. After analyzing the digital twin, the engineers can decide if it would need minor repair or replacement, thus increasing the process's efficiency and reducing time wastage.

10 Artificial Intelligence

Artificial intelligence, profoundly known as A.I. (artificial intelligence), is innovating sustainable production systems. Companies worldwide have started heavy investment to implement and improve the technology to better use to improve efficiency. The A.I. provides businesses with the considerable power of processing at an upscale level compared to human minds. The technology is also combined with machine learning ML to increase supply chain efficiency. The advanced processing technology will be trending to improve sustainable production practices in the future and in time to come. The A.I. helps companies to forecast consumer demand better and streamline production by automating the manufacturing scale. The vast data requirement for the A.I. processing stratified by the advent of 5G technology in I.C.T. In coming years sustainability targets will quickly achieve by businesses, as A.I. will monitor carbon footprints and wastages during the production process. The technology will optimize its inventory by correctly calculating the P.O.S. (point of sales) in real-time for supply chain benefit. The A.I. will enable companies to avoid surplus finished goods, preventing the triple bottom line's impact. The technology will indeed become part of all operations in the companies making them highly agile and lean at the same time.

Artificial intelligence that simulates the human brain helps companies like Excel Energy produce electricity by bringing coal to reduce Nitrous oxide emissions. Tech Giant Google reached a threshold point wherein data center efficiency reached its peak. It utilized A.I. models to increase data management efficiency when the algorithms in A.I. models reduce energy usage up to 40% for cooling data centers to impact its triple bottom line directly. Another tech giant I.B.M. uses A.I. to improve weather forecasting models for their solar energy production, 30% better sales demand, and manage energy loads to reduce greenhouse gas impact on the environment.

11 Industrial Revolution (I.R.) 4.0

The I.R. 4.0 is also an upcoming trend in sustainable production practices. The concept provides a framework that will reduce operating costs. The I.R. 4.0 provides complete digital internal and external integration in the supply chain. Information sharing is becoming a reality for companies adopting the practice. However, the cost to implement is high, but the benefits are much higher. Companies are trying to implement better customer services, optimizing production systems and their supply chains. Also, it provides excellent benefits in other sustainable production practices like recycling and remanufactures. The I.R. 4.0 is a framework to digitally communicate with machines, thus enabling smart manufacturing systems to reduce overproduction and energy consumption. Intelligent manufacturing systems are still in the infancy and are improving as time passes. Using the I.R. 4.0, the companies will help balance a company's resources, for instance, machines and workforce. Companies can also use the framework to have advanced sensors, controls, and tracking systems to avoid any untoward accidents because of hazardous chemicals, as witnessed in Lebanon's port explosion (WHO 2020).

12 Greener Energy Production

Greener energy production is now possible by using renewable energy sources like wind, water, and sun. Companies to achieve sustainable production practices invest massive capital in reducing carbon footprint, and eventually reaching carbon neutrality is not a distant future. The trends are picking up pace not only by companies but also by homes and schools by microgeneration. Green energy is not only clean but is economically viable in the longer run. Scientist has improved the capacity to store energy by producing efficient batteries to complement green energy production. Western and Asian companies have now embarked on making their business facilities carbon neutral. Companies are setting up solar and wind parks to harness clean energy. The technology to produce clean power is getting better and most efficient with the help of lightweight nanomaterials. The grids to distribute power for manufacturing consumption are becoming smarter due to A.I. and I.R. 4.0. With 5G I.C.T, things are picking up the pace to become more sustainable and efficient. The future trend will significantly reduce fossil fuel (coal, oil, and gas) consumption to halt the impacts of global warming and climate change.

13 Sustainable Commitments

Sustainable commitments are also an upcoming trend. The business across the globe are communicating their sustainable objectives as part of sustainable production practices. The direction will become part of many Asian countries. The European companies, however, due to regulation, are forced to reveal their initiatives for sustainable development in their business operations and the C.S.R. activities. The business in other parts of the world now follows the same trend to disclose practical steps for sustainable targets. Many companies are committing to reduce their carbon footprint and eventually become carbon neutral zero-till 2050 as per the vision of U.N. S.D.G.s. The reduction in carbon footprint will limit the global average temperature increase to 1.5 °C. The other bold commitments which will trend will be a green global supply chain run by clean energy and transportation. Companies are also becoming bold to publish carbon emissions during production publically on websites. The trend will help companies more customer loyalty, profits, and sustainable production practices in the future.

14 Carbon Negative the New Trend

The carbon negative is a new trend that is being embarrassed in sustainable production practices. Like bold commitments by companies, the carbon negative is also a significant announcement by companies worldwide. Companies like Microsoft, Walt Disney, Coco-Cola, and ArcelorMittal are taking pride in announcing their sustainable production practices by becoming carbon negative till the year 2030. The companies are pumping money in R&D to become carbon negative. Scientists and engineers are working around the clock to achieve the target with the help of new technologies. The initiatives by a leading name in different sectors will continue to follow by the rest of the global business communities.

15 Sustainable Logistics

Logistics plays a vital role in sustainable production practices; according to Chopra and Meindl (2013), logistics envelops transportation, storage areas, and inventory of a company. Transportation plays a massive part for companies to achieve sustainability in production. The energy used to manage and run operations in storage areas also significantly contributes to the cause. We have discussed in great length about raw materials and finished goods contribution to sustainable production practices in earlier stated trends. However, the focus for sustainable logistics is more toward the transportation and storage facilities in business companies.

The business enterprises will reduce their carbon footprint in logistics for environmental impact. The emission of carbon other effluents in the air impacts the triple bottom line of the company. The trend will become more permanent as governments worldwide have enacted regulations to force companies to bring better change to make logistics greener and efficient. Governments also have a role to play; transportation infrastructure in developing countries has to improve to provide cleaner options for companies to adopt. For instance, trains emit less pollution than trucks to move per ton of goods from one place to another. Besides freight options available for companies, the logistic managers also have to optimize their container loading to avoid any space, increasing logistic efficiency. The A.I. will make logistics more sustainable in production. The technology will enable to maintain the transport better, use effective routes for less fuel consumption and vehicle tracing to avoid any pilferage, and increase their road efficiencies to impact its triple bottom line. Third-party logistics (3PL) roles will increase in the coming days to provide smart and cleaner logistics and supply chain management options.

16 Corporate Digital Responsibility

Lastly, corporate digital responsibility will become a trend for companies to follow sustainable production practices. Corporate digital responsibility promotes ethical practices revolving around companies becoming digitally responsible. The responsible digital companies of the future will implement management to protect customer and employee data. The A.I. and other technicians who will be interpreting consumer behavior for sustainable production practices will flood companies with the massive data transforming into information that could harm customer and employee privacy. Digital responsibility also encompasses the economic aspect. Business organizations have to trade between employing technology (robots) and human resources to be economically digitally responsible. The pace is fast to replace humans for labor, but it will render many jobless and eventually to more hunger worldwide if machines would take over. The companies need to be digitally responsible for employing such technologies as it will go against the 2030 U.N. S.D.G.s, which ensure 'Zero Hunger' across the globe. The trend of digital responsibility will provide prudence in employing technology or the nature of the job, which increases the difficulty for human labor. The concept of digital responsibility in sustainable production practices will make companies rethink how to dispose of, recycle or reuse e-waste, and issues like energy consumption for data mining (bitcoin and big data).

17 Conclusion

The new era unfolds a wave of innovations and radical thinking, enabling activism to bring positive societal change. Governments, business organizations, and consumers are more aware of the accelerating environmental degradation. The pressure is mounting on all stakeholders to embrace sustainability in all walks of life. The most drastic change is needed to target UN SGD's by applying new trends in sustainable production. The organizations have to become responsible for choosing the systems of production. More objective-based tracking systems are required to check carbon emissions and wastages during the production process. The wisdom needed to overcome the triple bottom line challenges needs a higher level of awareness. Knowledge management is direly needed to remove any weak-force interactions in design and development. Improving sustainable production practices can happen by changing our focus from waste elimination to critical thinking on design flows and performances to achieve sustainability performance. With the help of knowledge management, the developed countries should reduce the gulf of sustainability development in developing and under-developed countries. A collective effort is a need of the hour to galop to reach the goal, which is crucial for all humankind and other species living on this beautiful planet called our home earth. Real sustainability can only achieve if we could better the lives of our future generations. The sustainability history we would make today will become a promising future for the next generation and a better future for ensuing generations to come. The right-thinking and trends of sustainable production practices are to set today, yielding better results tomorrow in the shape of a clean environment for all species to thrive and humans to progress.

References

Brosowsky A (2019) pMD Blog–5G: what does it mean for health care? https://Www.Pmd.Com/. https://www.pmd.com/blog/post/5g-what-does-it-mean-for-health-care

Chopra S, Meindl P (2013) Strategy, planning, and operation

Helen WP, Stefan SP, Joseph SP., Robert KP (2014) Sustainable operations management: recent trends and future directions. Int J Operat Prod Manag 34(5). https://doi.org/10.1108/IJOPM-12-2013-0557

Lees F (2012) Lees' loss prevention in the process industries: hazard identification, assessment and control. Butterworth-Heinemann

Pickup O (2020). 5G's environmental impact: can it ever be sustainable? https://www.Raconteur.Net. https://www.raconteur.net/technology/5g/5g-environmental-impact/

Shi W, Chen H, Zhang X, Ma C (2021) Optimal alliance strategy and its impact in a closed-loop supply chain by considering greenness and service effort. Math Probl Eng 2021:6676482. https://doi.org/10.1155/2021/6676482
Sustainability Guide (2018). Manufacturing–sustainability guide. https://Sustainabilityguide.Eu/. https://sustainabilityguide.eu/ecodesign/manufacturing/
United Nation (2015) 17 goals to transform the world for persons with disabilities. https://www.un.org/development/desa/disabilities/envision2030.html
United Nations (2020) Growing at a slower pace, world population is expected to reach 9.7 billion in 2050 and could peak at nearly 11 billion around 2100. UN DESA, United Nations Department of Economic and Social Affairs. https://www.un.org/development/desa/en/news/population/world-population-prospects-2019.html
Weizsäcker EU, Lovins AB, Lovins LH (1998) Factor four: doubling wealth-halving resource use: the new report to the Club of Rome. Earthscan
WHO (2020) Lebanon explosion 2020. https://www.who.int/emergencies/funding/appeals/lebanon-explosion-2020
World Commission on Environment and Development, Brundtland GH (1987) Presentation of the Report of the World Commission on Environment and Development to the Commission of the European Communities, the E.C. and EFTA Countries... 5 May 1987, Brussels. World Commission on Environment and Development.
World Nuclear Org (2020) Key role for nuclear energy in global biodiversity conservation. Blackwell Publishing Inc.

Responsible Production and Consumption Goals: A Fundamental Driving Forces of Economic Growth in Pakistan

Kiran Jameel, Saifullah, and Muhammad Asim Rafiq

Abstract The global significance of responsible production and consumption has gained greater prominence specifically in the context of developing economies like Pakistan where 25% of the overall population living below the poverty line, one-third of the population is struggling hard to eat three times a day but failed. Despite this, one-third of overall food production is being wasted globally. Besides many challenges, the consistent growth in urbanization, the transformation of lifestyle into modernization, and industrial re-evaluation are considered significant foundations. However, the current linear economic system, based on metal extraction, used in manufacturing, consumes the product and ends up disposing of it, which caused the world, particularly developing nations like Pakistan, to environmental problems, that ultimately reduce the biodiversity, air soil water, and pollution. Besides, to jeopardize the life on Earth support system, to reinforce and maintain the soil in terms of land-filling. Certainly, this lack of resources is a significant obstacle to achieve sustainable development goals and meeting the growing needs of the population that energies the economic growth of the nation. This chapter will describes and considers the prospective benefits, effects as well as contributions in achieving already sets goals in SDG 12—Sustainable Consumption and Production (SCP) on each sector of the economy especially in the context of developing nations like Pakistan.

Keywords Environmental degradation · Sustainable development goals · Clean energy · Economic growth

1 Introduction

Responsible production and consumption seek to reduce the destructive impression of production and consumption progression on the environment. While all aspects of the product and life service cycle supporting the quality of life as a whole. From a broader perspective, many operational strategies related to production and consumption are

K. Jameel (✉) · Saifullah · M. A. Rafiq
Hamdard Institute of Management Sciences, Karachi, Pakistan
e-mail: kiranjuw88@gmail.com

C. J. Chiappetta Jabbour and S. A. R. Khan (eds.), *Sustainable Production and Consumption Systems*, Industrial Ecology,
https://doi.org/10.1007/978-981-16-4760-4_7

central to the design and execution of fiscal, social, and environmental policies and initiatives, But their harmful contribution to the environment impedes the concept of sustainable development goals (Kimuli et al. 2021).

The nature conservancy plan has been implementing a constructive plan towards production and try to reconcile responsible production with conservation of biodiversity and social development that ultimately contribute to economic growth. Subsequently, the consumer's market has increased the demand 0f responsible production, and farming in Pakistan has also expanded to meet the required needs. Because of underlying reasons, the agriculture sector has improved its environmental performance that constructing more sustainability and competitiveness in the commodity market as a whole (Tseng et al. 2018). For environment conservancy, the successful stage is that when agriculture becomes fully sustainable and moves away from its old historical extensive production to a more efficient process of production that would contribute towards a better environment which reduced the amount of CO_2 emission (Steensland and Zeigler 2020). Intensifying sustainability in agriculture will not only be a good approach for the natural ecosystem but also increase the levels of production as well as increase the income level of farmers, employment in this sector, and the health of other life-supporting ecosystems (Hazemba and Halog 2021). Besides, sustainability is not concerned only with the agricultural sector but the challenges are hydra-headed the core challenges include sustainable development/growth and environmental preservation (Dogan et al. 2019). Even the most popular concept of sustainability is still a convincing idea "Sustainable growth is a growth which fulfills the essentials of the present generation without conceding the ability of future generations to meet their own needs (Heras et al. 2021).

To protect the environment from degradation and to achieve sustainable growth in the world where the human population is expected to reach 9.7 billion by 2050 (Shelden and Roessner 2013). Demonstrating that an additional burden of 2 billion persons to feed in the next few years, Urbanization, consumption habits as well as a change in living standards have contributed to the increasing food waste problem (Priefer et al. 2016) tackling these wasteful levels of production and consumption is essential. To ensure sustainable socio-economic prosperity, economic growth that would need to be disentangled from resource utilization and environmental degradation (Fatima et al. 2020).

This chapter described and considers the prospective benefits, effects as well as contributions in achieving already sets goals in SDG 12—Sustainable Consumption and Production (SCP) on each sector of the economy especially in the context of developing nations like Pakistan. Whereas SCP is also inclusive part of International Policy disclosure (IPD) for the last four decades yet the uptakes are still not smooth in Asia. Because some imperative socio-economic elements make SCP especially important in the Asia–pacific region. The earlier few decades have experienced the increasing trend of population, declining average income as well as enhance in starvation, urbanization and massive growth in the production activities. To fulfill the need of the growing population, to combat urbanization as a result of migration from rural to major urban cities, to minimize the waste of all-natural resources, to control

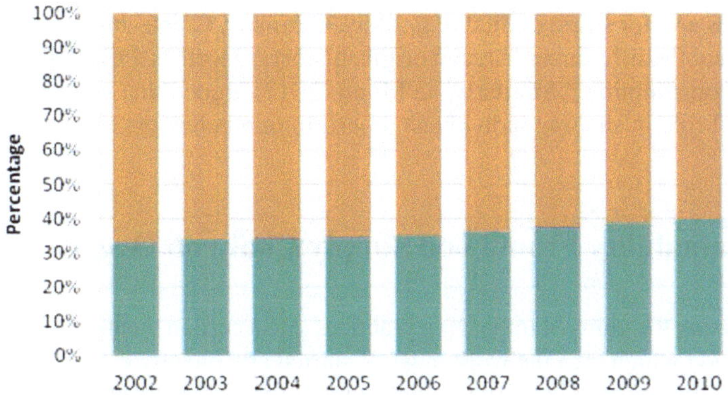

Fig. 1 Share of global manufacturing value-added, Asia–Pacific, and the Rest of the world. *Source* World Bank

the extraction of undersurface resources, to protect the environment from degradation the SCP is vital for current as well as the future generation. To aver come to the issue of starvation, urbanization, unemployment, and to satisfy the living standard of society as well as intensification in the business profit the business community increases the manufacturing of goods which results in environmental degradation. All the above-cited trends are meticulously 'coupled 'to inclusion in resource usage as well as emission. Figure 1 highlight the recent rise in manufacturing across the region.

However, different elements of micro and macro level can describe cross-country changes in sustainable consumption (SC). The SC is carved collaboration with the individual lifestyle and with the context of national preferences (Thøgersen 2017). The National culture of any community is viewed as the most influential macro factor that determines sustainable consumption. In his chapter on the elements describing cross-national changes in organic food consumption, Thøgersen (2017, p. 171) argues, "Macroelements likewise food culture as well as the culture's level for post-materialism, environmental issues play a substantial role." He further emphasizes the significance at the level of political regulation, the inclusion of legal definitions, set up of updated standards, farmers support in monetary terms, the national cataloging system as well as the soil conditions. As SC decision-making is complex, so the list of the potential factors does not finale yet.

The level of environmental consciousness, dissemination substructure, and income level also include extra components considering the cross-country variant (Milfont and Markowitz 2016; Nair and Little 2016; Thøgersen 2017). To address this complexity, confirmation about the impact of accessibility of product choices on SC is diverse (Hemmerling et al. 2015).

To conclude that macro or micro-element can fully enlighten the existence of differences in SC across the countries (Milfont and Markowitz 2016). However, the spotlight on the role of cultural effects is permissible. Nair and Little (2016) argue that national culture is a considerable critical factor: "Obviously in those countries where the cultures are not strictly allied (e.g., Italy, Germany, China, Japan, Malaysia, and India) the subtle differences in racial origin, history, as well as the religion, will affect on green consumption." Similarly, de Mooij (2017) argued that cultural differences play a vital role in shaping individuals' consumption behaviors.

2 Sustainability From Food Supply Chain an Overview

After realizing the gains of improving practices into a certain location it is also important to decide which area the work must be started and the process must be changed. Expanding the services and production efforts into underutilized areas have high or medium potential for agriculture and have been deforested already can lead to value in sustainable production (Steensland and Zeigler 2020). The example of Brazilian agriculture is embedded wherein the efforts and innovative systems of Embrapa are implanted in the Brazilian agriculture environment that is both competitive and successful (Parente et al. 2020). Even in times of severe economic turndown like the period of 2014–2017, where people were losing income and the country was not stable economically, this sector and its sustainable ways have been the source of income, jobs, food security for the population. The sectors also had spare production for exports that would eventually help the country come out of the economic turmoil drawing a surplus in the balance of trade (Parente et al. 2020). Currently, the agriculture sector alone is contributing 25% to the GDP, and not only that but a major contributor to overall employment at the rate of 40 to 50%. In 2016 the country faced some climatic constraints particularly in the areas where the important product is grown and there was a reduced grain production, however, due to the efficient process in the 2016–2017 cycle of production reached the record grain harvest amounting to 240 million tons (Ministry of Planning Pakistan, 2019a, b). This is only precedent from developing nations but it was the first stone which was a throne. The continuous innovation in this sector has proven to the key to success in the agriculture sector. Embrapa has the factors which have played a major role in bringing innovative ways to improve the sustainability and efficiency of this sector hence making farming reach the top level of success. At this point, there is a presence of an extremely strong private agriculture sector with several local and renowned multinational companies that are guiding and providing interminable options to the farmer community (Gul et al. 2018). The job of Embrapa is to make strategies and agendas which are many times in partnership with the private sector so that the results can be enhanced and the innovative solution could be reached to make the sector more competitive than before as well as more productive. The sixth Embrapa Master Plan (Embrapa 2015) recommends that the strategies should be such that increase the productive competitiveness alongside sustainable development in the sector (Ghaus et al. 2016).

3 Sustainable Development Goals (SDGs) 2030

The United Nations (UN) has adopted the Sustainable Development Plan 2030 (SDP) in September 2015 to achieve 17 Sustainable Development Goals (SDGs) at the end of 2030 (Griggs et al. 2013) accompanying their 232 allied indicators (Key indicator for Asia and pacific 2019) by the member of UNO countries. The Asian countries are also included in the list that has dedicated themselves nationally to adopt the agenda of 2030 as well as to meet the priorities and goals (Cho et al. 2016). Resource and energy conservation, sustainable development and to access better employment opportunities as well as standards in quality of life can be accomplished through sustainable production and consumption (Gasper et al. 2019). Sustainable growth is a center of change in building a sustainable economy that intensifies on a distant future basis both for the world as well as for the populace (UNDP 2016). To achieve prosperity global and sustainable development standards the way is only to reform/ regenerate or recycle used goods and services that can yield the least waste by this way we can reduce our environmental effect. The agriculture sector is the world's greatest water user, and agriculture today is consuming about 70% of all available freshwater which is fit for human usage (Steensland and Zeigler 2020).

On the other hand, a substantial percentage of the world's population is struggling hard to eat but minimum to satisfy their hunger needs. Half of the worldwide food waste is generated at the level of retailers and customers that is also critical for attaining an efficient and effective food supply chain, by reducing the food waste which will help in food security as well as move us into a more capital productive economy (Priefer et al. 2016). To accomplish this aim, the effective use of available natural resources and the way people disposing of hazardous waste and toxins are critical goals. It is equally vital to facilitate enterprises, companies, and customers to recycle and minimize waste, as well as to help developing countries transition towards more sustainable consumption habits by 2030 (Parente et al. 2020).

Presently, about one-third of overall food production is discarded or destroyed on the planet by the people, whereas almost 690 million peoples are hungry (Gasper et al. 2019). Producers and cultivate agricultural products are required to produce more food to feed the planet sustainably considering reduced detrimental environmental effects such as soil, water, and nutrient depletion, greenhouse gas emissions, and habitat destruction. Consumers with a diminished environmental impact must be persuaded to move and avail healthy and superior nutrition. The decrease of natural supplies and growing urbanization, as well as the rising trend of world population, means feeding more people with less water and other allied agriculture products (Sikdar 2011). To satisfy the anticipated increased water demand, electricity and food are required to move towards more efficient approaches to production and consumption (Jiang et al. 2012).

The Sustainable Consumption and Production (SCP) in Pakistan the National Action Plan (NAP) on Sustainable Development Goal 12 (SDG12) was established after conducting a comprehensive study as well as with the consultation of all stakeholders. According to Churchman (1967) study a prosperous environment, responsible production and consumption are vital. But by the use of a linear economic system, human and environmental processes commensurate in different forms which emerge a lot of unsustainable problems. It is a non-trivial exercise to address these issues and may be considered one of the "wicked" challenges for the planet. Impressive issues are often unexpected and have an inadvertent impact, dynamic, intractable, contradictory, and multidimensional problems. However, the use of soft computing from a responsible production and consumption viewpoint is a powerful collection method to cope up with wicked problems.

Similarly, Griggs et al. (2013) reported that Sustainable Development Goals (SDGs) set by United Nations was established to set an agenda for transforming the industries, economies, and cultural habits transition towards more sustainable by 2030. An integral part of the Global Growth Agenda before 2030, sustainable production and consumption had been acknowledged. It is separately defined as an objective at 12 numbers and, at the same time concerning some other accepted objectives.

The Vision 2025 model was implemented which is also based on sustainable usage of production concepts (Gasper et al. 2019). It intends to put Pakistan on a pathway to steady growth with the ultimate goal of ranking it one of the world's top twenty-five economies by 2025 as well as the top ten by 2047. Several fundamental driving forces are vital for achieving responsible production and consumption for developing economics and for the economic growth of the country (Ghaus et al. 2016).

4 Sustainable and Affordable Clean Energy Production

The collective efforts of every researcher to keep the efforts productive as possible, a corporal aspect recommend is to make a team mostly consist of youngsters, launch an app to collect all possible data about garbage, industrial waste as well as collect other waste caused by the means of transportation from cities to country. Finally, the collected garbage can be recycled and waste is producing affordable and clean energy through biogas technology. Secondly, another alternative viably and the cheaper energy source is wind and solar power which is a comparatively less costly and clean source of energy having the least climate impact also known as a natural source of energy. Undoubtedly, the Technology Need Assessment (TNA) is trying its best to determine the best possible technologies for producing a renewable energy transition to minimize pollution. The planned actions would minimize the uncertainties associated with the impact of environmental degradation. It would also be built a resistance against disasters caused by environmental degradation and secure the ecosystem, land as well as economy. Resultantly the country's development direction will move to optimal level by attaining a low carbon road, it would contribute to the international initiative for environmental control and its degradation.

5 Sustainable Innovative Infrastructure and Industry

Sustainable economic growth in Pakistan is facing a significant challenge owing to the scarcity of advanced technology, promoting in-fracture, trained and qualified manpower. These shortcomings have a significant impact on countries' economies with high degree emissions, leakage from air, water, land, and environment degradation result in untreated wastes. Moreover, the manufacturing sector is not in a position to maintain sustainable development due to low spending as well as a lack of research and development (R&D). There is also a lack of knowledge and resource sufficiency in the industrial sector to adopt sustainable development innovations and to increase awareness about environmental degradation and potential financial benefits associated with them. This technological and other allied industries insufficiencies generate favorable opportunities for volunteer investors in collaboration with experienced entrepreneurs to evaluate the basic requirements of the country that is being filled by imports. Inductee them inside the country, it will bring an economic revolution in international trade. Most importantly it will raise the high quality of education in terms of new technology and innovation in the field of sustainability.

6 Prospectus of Communities and Cities in a Sustainable Environment

To laid urbanization on a significant growth regime, it needs to transform the cities into the most innovative, eco-friendly as well as sustainable communities through the advanced urban planning system, reliable connectivity for local mobility (Mass Transit systems), better urban government and improve the public protections. A shift towards the vertical growth in city centers would provide the residents advanced residential facilities in addition to economic space. Putting the communities and cities on the path of smart communities'-cities will provide urban growth as well as capable to respond the rising complexities and demand of information technology. Make sure that Pakistan's cities are digitally strengthened, fitted with wireless network devices wherein e-connectivity is accessible in all places that will help in the free flow of information. Public transport, like rapid rail networks, will be carefully planned and introduced to cut down on the use of private transport in urban centers. Besides, the processes for the treatment of patient waste would be strengthened to have a reduced environmental impact.

7 Sustainable Environment Leads to Women Empowerment and Gender Equality

To make the physical environment and community peaceful, respectful and acceptable for all race/genders to live need to Promote and grow the notion of sustainability. In the broader concept of humanity, there is no competition between both genders in terms of respect and rights both have the same computable standards, even for women it's more suitable, despite all, dignity needs to be maintained by both for each other. Ensure equal opportunities to economic growth, equal rights to access social welfare, right of ownership, land management and all types of other property, equal rights in inheritance, natural resources, right to access emerging technologies and financial services including microfinance both for men and women especially to the poor and deprived people of the society. Similarly, Nour (2009), pointed out a social injustice with feminine and argue that early-age marriage can threaten a girl or young woman's educational performance. It may affect their prospects for employment; the form, other hidden circumstances for their future employment conditions; their physical well-being as well as the welfare of their pro-genies.

8 Clean Water and Sanitation Facility

Nowadays, limited access to clean water is declared as one of the most imperative issues owing to rapid growth in the world population which consequently increased food supply and water demand, social life as well as industry (Shannon et al. 2010). Issues concerning water were also called out by a large number of scholars and academia to put a maximum consideration on water sustainability as well as to pay utmost effort in discovering robust technologies for the treatment of wastewater and desalination besides in improving the productivity of available existing water production as well as its distribution systems (Sikdar 2011). Almost a lot of potential solutions such as film procedures like nan-filtration (NF), microfiltration (MF), ultrafiltration (UF) and reverse osmosis (RO) were established for devastating implementation in the water industry. But these available technologies are energetically or chemically exhaustive consequently excoriated due to the highest cost, extensive exergy, and chemical consumption, and much muddier tendency that involves frequent cleaning or backwash. Forward osmosis (FO), employing natural phenomenon of osmosis that is a newly evolving membrane technique determined by osmotic gravity incline generated through a semipermeable membrane with two flowing streams of varying concentration (i.e., the draw solution and the feed). So, the exergy needed to throw water across the membrane is almost negligible. Far from being so, FO generates a little problem of muddier as well as its cleaning (Mi and Elimelech 2010). Due to having these exceptional features, FO differentiates itself from other available membrane procedures for the supply of sustainable clean water.

Audrey Hepburn Quote in the twentieth century "Water is life, and clean water means health". Hence, life is an utmost priority from all physical aspects. For better living standards and good health availability of clean water to the entire world, the population is essential. Besides all available advanced clean water techniques still in 2020, no serious effort has been made to arrange clean water availability and its proper sanitization facilities. Although enough freshwater is available on the planet for supply. Yet, millions of people expired with diseases that are directly associated with the usage of unclean/ unhealthy water every year. To mitigate the risk and prevent the population from diseases associated with the usage of unhealthy water, vigilant eyes from water management are essentials.

The above Figure shows that out of 25 regional economies just 08 countries enjoying the clean and fresh water facility which surpasses 90% as per available data of 2017 (see Fig. 2). This proportion hit the figure of 100% in the case of Hong Kong, China; New Zealand; and Singapore are within these 8 economies that utilizing 90%. Conversely, in 9 other economies, the percentage of people that consumed clean drinking water facilities was below 50%. Since the year 2000, development has been made to deliver secure and controlled drinking water to all but 3out of the 25 reporting economies. Similarly, Food wellbeing, secure living options, and educational opportunities may be adversely impacted by inadequate sanitation. As per available data for the year 2017 reflect that out of 14 regional economies 90% population enjoying the facility by using sanitation that was operated safely observed only in 04 regions. Whereas the ratio in Singapore is 100%, the Republic of Korea 99.9%, and Japan 98.8 with the coverage of such services up to 100%.

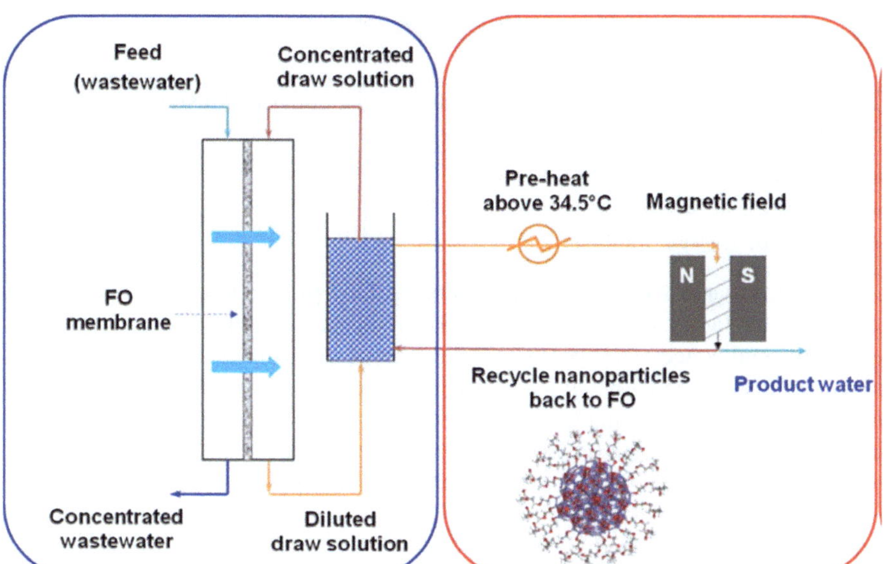

Fig. 2 Wastewater and recycle nanoparticles back to FO. (*Source* Su et al. 2012)

9 Environmental Sustainable in the Context of Asia

wastage is not only an issue for developing countries, but a global problem related to food safety and protection, and has other important sustainability aspects. Food production and unhealthy consumption lead to a waste generation that has several causes, some from food producers and retailers, which can be grouped into three major categories, i.e. natural constraints, mega-trends, and management. The food waste issue has a rigorous impact on three main pillars of sustainability:

I. **Economic Impact**
Food waste has an overall impact on the economy, such as significant crop losses during the processing or storage of food items. The economic impact can be calculated via the life-cycle costing technique (LCC).

II. **Social Impact**
Growing poverty in developing/low-income countries.

III. **Environmental Impact**

Nearly 14,000 million hectares of productive land (28% of the cultivated region of the earth). They are used annually to produce enough food which is wasted or destroyed. It is estimated that The carbon emissions of wasted food contribute to greenhouse gas (GHG) pollution by adding around 3.3 billion tonnes of CO_2.

9.1 Assessment of Environment Impact

The environmental effects of food waste can be measured using the following techniques:

- Life cycle Analysis
- Material flow Analysis
- Energy or Exergy Analysis

The term that is widely used in sustainable food production is organic farming, which Walter James said in his book "Look at the Land" in the 1940s. The topic of environmental concern was discussed in this book on chemical farming. After this, sustainable performance has gained value in the industry associated with food production. Subsequently, sustainable performance has gained value in the industry associated with food production (O'Brien 1999). While the idea of sustainable production was coined in the year 1992 at UNO in the Conference on Environment and Development, which is a core element of sustainable development, academics, scholars, and researchers have described sustainable production in a variety of ways. However, this term is well established by Lowell Center for Sustainable Production (1998) (LCSP).

- Generation of goods and services by this way that has non-polluting
- Protecting natural resources and energy
- Healthful and safe for communities, consumers as well as employees
- Economic viability
- Creatively rewarding to all such as society and people (Glavič and Lukman 2007; Tseng et al. 2018)

Similarly, the American Public Health Association (APHA 2007) describes "one which delivers nutritious foodstuff to fulfill existing food needs and preserving healthy ecologies that can also deliver food for generations to come with least adverse ecological impacts. Sustainable food system also supports domestic production and distribution infrastructure and makes healthy food available, accessible, and affordable to everyone." It is inferred from the concept that sustainable production goals can be accomplished by healthily preserving the current ecosystem, minimizing waste that can protect the environment from degradation. The Sustainable Development Goals (SDGs) adopted by the Members of the United States cover both sustainable food production and consumption aspects within their established objectives. The Life Cycle Assessment (LCA) methods are well adapted for the promotion of these objectives. These techniques cover not only the stage of the food chain but also upstream, i.e. the processing of raw materials, animals, agriculture, fisheries, and fish farming, as well as the packaging of products. The downstream covers the distribution of goods, consumption, and disposal/waste also covers transport activities at all levels.

To mitigate severe environmental effects plant greenery that will lead and improve the percentage of forestation within a country so that favorable physical environmental changes will occur. The only way to fight the worst environmental situation is to eliminate pollution in our daily routines. Further, we have to reduce the level of carbon emission through innovative as well as environmentally friendly manufacturing techniques. Sound calamity risk reduction measures are being developed by the Asian Pacific region.

Environmental change is a distressing global challenge, considering its small exposure of 0. 88% to global GHG pollution, Pakistan ranks eighth in the list of countries having the most vulnerable impact on climate change. The most serious threat to Pakistan in the future is climate change caused by increased instability in the monsoon season, as well as floods and droughts. In the case of arid and semi-arid areas, extreme water and heat stress conditions caused the reduction of agricultural production and power generation, resulting in upstream saline water intrusion in the Indus Delta. This will adversely affect the coastal crops, mangroves, and fish breeding grounds in coastal regions, including the cities of Karachi, Thatta, and Badin, which are also in threaten by increasing sea levels and intensified cyclonic activity. The agricultural production, water shortages, mangroves, and fish farms as well as a threat to marine areas are seriously impacted by this climate change. Therefore, climate change issues are explicitly and badly impacting the plans in Pakistan for environmental sustainability.

10 Sustainable Agricultural Productivity

One of the most imperative issues for Pakistan is food security. The vital element of Pakistan's economy is agriculture, which is based on food security as well as the source of raw materials for various industries and have significant exports. Achieve food security in Pakistan will remain a persistent problem, despite substantial economic growth and a major rise in agricultural production. The affecting factors are inefficiencies in food distribution, rising population growth, prospects of climate change, depleting water supplies, least investment in agricultural research and development (R&D) as well as unsatisfactory food safety which is a net for extreme poverty. The state of hunger remains serious in the world as well as in Pakistan that ranks 93 out of 104 countries in the Global Hunger Index (GHI) ratings. Pakistan which was one of the wheat exporter countries is now in danger and struggling to satisfy its domestic wheat demand. Owing to inequality uses of fertilizer products decreased water supply, low seed quality, inefficient and bad quality of farm equipment, the average production per unit is also low protections through the application of climate-resilient techniques. The Sustainable Food Action Plan has four key objectives: to concentrate on the productive agriculture sector, to eliminate food waste, to conserve agricultural land, and to ensure food security through the implementation of climate-resilient techniques. The Action Plan also provides a timeline for addressing the related SDG goals prioritized by SCP. This proposal stresses the change in the food supply chain toward a more balanced food environment. Pakistan needs strategic planning to enhance agriculture and food security, which ensures structural improvements in the way food is produced, stored, transported, and consumed. This will change aggregate trends of consumption and development and will help to maintain the country's food security. To reflect the factual position we obtain a Figure from ILOSTAT which is given in Fig. 1.

In 2017, out of 37 reporting countries, 26 are the incidence of undernourishment in the overall population. The prevalence of obesity in Azerbaijan, Australia, Hong Kong, Japan, Kazakhstan, China, New Zealand Malaysia, and Korea was the lowest, at a rate of 2.5% or below. Furthermore, Afghanistan (29.8%), Timor-Leste (24.9%), and Pakistan have seen the highest rates of undernourishment in 2017 (20.3%). As shown in the above-mentioned figure.

11 Conclusion

This book chapter has tried to cover all aspects relating to responsible production and consumption (SCP). A detailed deliberation has also been made on the environmental degradation caused by waste generation from different aspects of consumption as well as people's attitudes. The overall worldwide ambition of responsible production and consumption is to reduce the destructive impression of production and consumption progression on the environment. As sustainability is a broader concept but this study has covered its different risk areas have been after realizing the gains of improving practices into a certain location it is also important to decide which area the work must be started and the process must be changed. Expanding the services and production efforts into underutilized areas have high or medium potential for agriculture and have been deforested already can lead to value in sustainable production. This chapter also covers the following sub-heading on sustainability concepts:

- An overview of the food supply chain,
- Sustainable Development Goals (SDGs) 2030
- Sustainable and affordable clean energy production,
- Sustainable as well as innovative infrastructure and industry,
- Prospectus of Communities and cities in a sustainable environment,
- A sustainable Environment leads to Women empowerment and gender equality,
- Clean water and sanitation Facility,
- Environmental Sustainable in the context of Asia and
- Sustainable agricultural productivity

In this chapter example of Brazilian agriculture is embedded wherein the efforts and innovative systems of Embrapa are implanted in the Brazilian agriculture environment that is both competitive and successful Even in the times of severe economic turndown like the period of 2014–2017, where people were losing income and the country was not stable economically, this sector and its sustainable ways have been the source of income, jobs, food security for the Brazilian population. The sectors also had spare production for exports that would eventually help the country come out of the economic turmoil drawing a surplus in the Brazilian balance of trade. Currently, the agriculture sector alone is contributing 25% to the GDP, and not only that but a major contributor to overall employment at the rate of 40–50%. While considering the context of Pakistan this study found that the vital element of Pakistan's economy is agriculture, which is based on food security as well as the source of raw materials for various industries and have significant exports. Achieving food security in Pakistan will remain a persistent problem, despite substantial economic growth and a major rise in agricultural production. The affecting factors are inefficiencies in food distribution, rising population growth, prospects of climate change, depleting water supplies, least investment in agricultural research and development (R&D) as well as unsatisfactory food safety which is a net for extreme poverty. The state of hunger remains serious in the world as well as in Pakistan that ranks 93 out of 104 countries in the Global Hunger Index (GHI) ratings. Pakistan which was one of the wheat exporter countries is now in danger and struggling to fulfill its domestic wheat demand. Owing to inequality uses of fertilizer products decreased water supply, low seed quality, inefficient and bad quality of farm equipment, the average production per unit is also low protections through the application of climate-resilient techniques. The Sustainable Food Action Plan has four key objectives: to concentrate on the productive agriculture sector, to eliminate food waste, to conserve agricultural land, and to ensure food security through the implementation of climate-resilient techniques. The Action Plan also provides a timeline for addressing the related SDG goals prioritized by SCP. Besides, some motivational factors have not been differently formed through vertical collectivism, as well as horizontal individualism.

12 Recommendation

To gain sustainable benefits from the food sector in developing nations like Pakistan we develop a book chapter on the topic based on the review, in-depth survey of available literature the authors recommend the following suggestions:

- Devise a 10-Year Framework Plan on Sustainable Consumption and Production (SCP) Patterns by all countries in the world and take action together with developing nations play a leading role by considering the progress as well as their capabilities available in Developing countries.
- The target or policies may be devised to achieve sustainable management for the efficient use of natural resources by 2030.
- Target may be set to reduce overall food wastage within-country levels and diminish food losses along production and supply chains, including post-harvest losses.
- Policies may be devised to adapt the Circular Economy concept which substantially reduces waste generation through prevention, reduction, recycling, and reuse.
- Incentive policies may be devised to encourage large-scale companies to implement sustainable practices within the organization.
- Endorse public procurement practices through Public Procurement Regulatory Authority (PPRA) which are more sustainable, following national policies and priorities.
- Peoples' awareness campaigns may be started everywhere to educate the people about sustainable development and lifestyles in harmony with nature.
- Devise and implement a system that monitors the sustainable development effects on sustainable tourism (ST) which generates jobs, upholds native culture as well as products.

References

Apha A (2007) WEF (2005) Standard methods for the examination of water and wastewater. American Public Health Association, American Water Works Association, and Water Environment Federation

Cho J, Isgut A, Tateno Y (2016) Pathways for adopting the sustainable development goals to the national context: the case of Pakistan. Asia Pac Dev J 24(2):53–87

Church Griggs D, Stafford-Smith M, Gaffney O, Rockström J, Öhman MC, Shyamsundar P, Noble I (2013) Sustainable development goals for people and planet. Nature 495(7441):305–307

Churchman CW (1967) Free for all. Manage Sci 14(4):B141–B142

de Mooij M (2017) Comparing dimensions of national culture for secondary analysis of consumer behavior data of different countries. Int Mark Rev
de Moura Oliveira J, Madari BE, de Melo Carvalho MT, Assis PCR, Silveira ALR, de Leles Lima M, de Almeida Machado PLO (2018) Integrated farming systems for improving soil carbon balance in the southern Amazon of Brazil. Reg Environ Change 18(1):105 116
Dogan E, Taspinar N, Gokmenoglu KK (2019) Determinants of ecological footprint in MINT countries. Energy Environ 30(6):1065–1086
Elia G, Margherita A (2018) Can we solve wicked problems? A conceptual framework and a collective intelligence system to support problem analysis and solution design for complex social issues. Technol Forecast Soc Chang 133:279–286
EMBRAPA (2015) Secretariat for Institutional Management and Development. VI Embrapa Master Plan 2014–2034. Brasília, DF, p 24
Fatima N, Ashraf I, Zehraa S (2020) Sustainable consumption and production: an analysis of implementation of Pakistan's policy framework of SDG-12
Gasper D, Shah A, Tankha S (2019) The framing of sustainable consumption and production in SDG-12. Global Pol 10(S1):83–95. https://doi.org/10.1111/1758-5899.12592
Ghaus K, Ahmed N, Toru SK, Manzoor R, Sohaib M (2016) Implications of implementing SDGs at the national level: a case of Pakistan (No. 32). Southern Voice Occasional Paper. https://ideas.repec.org/p/svo/opaper/32.html
Glavič P, Lukman R (2007) Review of sustainability terms and their definitions. J Clean Prod 15(18):1875–1885
Gul M, Gul E, Mantegazza I (2018). Measuring sustainable development in Pakistan data ecosystem for agenda 2030. Nottingham
Hazemba M, Halog A (2021). Systematic review of how environmental management policies are incorporated into national development plans in order to achieve sustainable development. Environ Challs 3
Hemmerling S, Hamm U, Spiller A (2015) Consumption behavior regarding organic food from a marketing perspective—a literature review. Org Agric 5(4):277–313
Heras A, Relinque-Medina F, Zamora-Polo F, Luque-Sendra A (2021) Analysis of the evolution of the sharing economy towards sustainability. Trends and transformations of the concept. J Clean Prod 291
Jiang J, Huo Z, Feng S, Zhang C (2012) Effect of irrigation amount and water salinity on water consumption and water productivity of spring wheat in Northwest China. Field Crops Res 137:78–88. ISSN 0378-4290, https://doi.org/10.1016/j.fcr.2012.08.019. (https://www.sciencedirect.com/science/article/pii/S037842901)
Kimuli JB, Baofeng D, Zhang R, Shaolin W, Jierui L, Yin W (2021) A multisource trend analysis of floods in Asia-Pacific 1990–2018: implications for climate change in sustainable development goals. Int J Disast Risk Reduct 59:1
McMillan C, Overall J (2016) Wicked problems: turning strategic management upside down. J Bus Strat
Mi B, Elimelech M (2010) Organic fouling of forward osmosis membranes: fouling reversibility and cleaning without chemical reagents. J Membr Sci 348(1–2):337–345
Milfont TL, Markowitz E (2016) Sustainable consumer behavior: a multilevel perspective. Curr Opin Psychol 10:112–117
Ministry of Planing (2019a) Pakistan implementation on sustainable development goals. https://www.pc.gov.pk/web/sdg/sdgpak
Ministry of Planing (2019b) Pakistan national action plan. https://www.pc.gov.pk/uploads/report/National_SDGs_Framework_-_NEC_2018.pdf
Nair SR, Little VJ (2016) Context, culture and green consumption: a new framework. J Int Consum Mark 28(3):169–184
Nour NM (2009) Child marriage: a silent health and human rights issue. Rev Obstet Gynecol 2(1):51

Parente R, Melo M, Andrews D, Kumaraswamy A, Vasconcelos F (2020) Public sector organizations and agricultural catch-up dilemma in emerging markets: The orchestrating role of Embrapa in Brazil. J Int Bus Stud 1–25
Priefer C, Jörissen J, Bräutigam KR (2016) Food waste prevention in Europe–a cause-driven approach to identify the most relevant leverage points for action. Resour Conserv Recycl 109:155–165
Shannon MA, Bohn PW, Elimelech M, Georgiadis JG, Marinas BJ, Mayes AM (2010) Science and technology for water purification in the coming decades. Nanosci Technol Collect Rev Nat J 337–346
Shelden MC, Roessner U (2013) Advances in functional genomics for investigating salinity stress tolerance mechanisms in cereals. Front Plant Sci 4:123
Sikdar SK (2011). What about industrial water sustainability? Clean Techn Environ Policy. Springer-2011
Steensland A, Zeigler M (2020) Productivity in agriculture for a sustainable future. In The innovation revolution in agriculture. Springer, Cham pp. 33–69
Su J, Zhang S, Ling MM, Chung TS (2012) Forward osmosis: an emerging technology for sustainable supply of clean water. Clean Technol Environ Policy 14(4):507–511
Thøgersen J (2017) Sustainable food consumption in the nexus between national context and private lifestyle: a multi-level study. Food Qual Prefer 55:16–25
Tseng ML, Zhu Q, Sarkis J, Chiu AS (2018). Responsible consumption and production (RCP) in corporate decision-making models using soft computation. Ind Manag Data Syst
UNDP (2016) United Nations Development Programme. One United Nations Plaza, New York, NY, p 10017

Achieving Sustainable Competitive Advantage Through Inventory Management Practices: The Case of Homegrown Coffee Business

Revenio C. Jalagat and Perfecto G. Aquino

Abstract This paper explores the effectiveness of the inventory management practices employed by a homegrown coffee business to attain profitable operations and sustain competitive advantage. Interview and focus group discussion was conducted to gather valuable information about their responses on the significance of inventory management in value creation, financial and operational performance, and to create competitive advantage. Key findings revealed ineffective use of inventory, ineffective use of materials, lack of emphasis on quality materials, inadequate demand forecasting resulting in either over or underproduction. Based on these findings, it is suggested to adopt effective and efficient inventory management practices, the need to reassess and review the current inventory management practices to determine areas of improvement. Moreover; the quality culture of production should be adopted and benchmarking with high-performing businesses in the industry. Lastly, management of the inventory should deal with the processes of maintaining, controlling, monitoring, and tracking inventory mobility from acquisition to final disposition to ensure the upkeep of optimum stock levels to meet production goals and customer demands.

Keywords Inventory management practices · Homegrown coffee house · Sustainability · Competitive advantage

1 Introduction

Considered as among the five essential commodities in the agriculture sector worldwide is the coffee product according to various authors (Gabriele and Vanzetti 2005; Ibrahim and Zailani 2010; Taylor 2005). Subsequently, coffee is among the reputable beverages which are consumed by people globally and thus, contributed positively

R. C. Jalagat (✉)
Al Zahra College for Women, Muscat City, Oman

P. G. Aquino
Duy Tan University, Danang City, Vietnam

C. J. Chiappetta Jabbour and S. A. R. Khan (eds.), *Sustainable Production and Consumption Systems*, Industrial Ecology,
https://doi.org/10.1007/978-981-16-4760-4_8

to the economy of the world. In the category of non-staple food, coffee product occupied the first rank in global trade and has also been ranks fifth as the most agricultural product. Kaplinsky (2004) stressed that the industry of coffee has been named as the second-highest commodity in terms of importance behind the oil which contributed to earnings of US$ 60 billion globally. Also, around 125 million people consider coffee as their main reference to living. The revenue and economic potentials of coffee production as a worldwide product have spurred different countries including the Philippines to raise and ameliorate the performance of the industry specifically in areas of how to manage inventory leading to attaining competitive advantage in the worldwide marketplace.

In the Philippine setting, quite several firms that belong to coffee production are encountering sustainability issues because of the lack of effective management, weak forecasting, scheduling issues, high cost of production, and lack of emphasis on procedures and processes. These factors impact negatively industry productivity that caused failure to attain a competitive advantage in the international arena. Moreover, most coffee farms in the Philippines do not possess effective inventory management practices that will help improve coffee production compared to other countries. With this, the business of homegrown coffee house chain which was first established in 1996 has been of those affected with this apportioned struggles. The business then grew into more than ninety (90) branches throughout the country since its first operation. As a privately owned firm and a market-based business, the production and the prices are controlled by the demand and supply of goods and services rather than the government.

Also, the main issues circulated were focused on the lack of provision by the homegrown coffee business of material sufficiency for production as well as the supply of products to consumers when needed and continuously. The recent practices on inventory management by homegrown coffee require realignment with what should be effective and efficient practices and strategies. Attaining these realignments will potentially resolve the demand and supply imbalance and eliminate inconsistencies in the production processes. Based on this context, firms necessitate re-assessment of their existing inventory management practices and devise means to develop a conceptual basis for attaining competitive advantage in the marketplace and devise ways to offer superior products and services to customers in line with the view of maximizing investment in inventory. To reiterate, this paper primarily aimed at investigating how the management inventory practices of homegrown coffee attain a competitive advantage in the marketplace in the Philippine setting. Moreover, it seeks to address the following issues? (1) Are the inventory management practices utilized by homegrown coffee houses effective? (2) What challenges do the homegrown coffee house faced in adopting the inventory management practices? (3) What are recommendations needed to improve the inventory management practices to achieve sustainable competitive advantage?

2 Literature Review

2.1 Coffee Production in Global Scale

The coffee industry has seen a tremendous change over the last 30 years and especially in developing nations, the market of coffee has been colorful and vibrant with differentiated coffee products such as eco-friendly, organic, and fair trade to foster the development of communities (Ponte 2002). The differentiation of coffee has practically helped the coffee producers or farmers maximized opportunities for larger production especially amid economic challenges in the coffee industry. Rigorous enforcement of standards was made by the government concerned authorities to ensure that control mechanisms are in place, social and environmental factors, the structure of pricing, as well as the inventory management practices that eventually lead to quality production of coffee products worldwide.

One of the profitable products in the world trade is coffee. It has catapulted the industry as the second top commodity traded globally. Ibrahim and Zailani (2010) argued that coffee production has promoted employment to millions of people at the international level from growing, processing, cultivating, and marketing of this commodity where it was seen as the core source of market trading in New York and London. Furthermore, the coffee product can be seen in over 80 countries globally especially the developing countries, and produced by 25 million farmers in Asia, Latin America, Africa occupied by the large majority of small-scale growers with less than 10 ha of land (Greser and Tickell 2002). These farmers have considered coffee farming as their main source of livelihood.

The evolution of coffee production can be recalled in the'60 s wherein the industry was supervised by the International Coffee Agreement (ICA) wherein they are responsible for setting the target price and assigned quota for exports of coffee to each country's producers. Using this system of pricing and quota allocation experienced challenges at the beginning stages, however; Ponte (2002) stressed that it later achieve success in stabilizing coffee prices until 1989 when the ICA regime has died as the inevitable change in the power balance in the supply chain of coffee globally. This has led to the liberalization of agricultural markets in coffee-producing countries and the decrease in intervention by the government in the marketing and export of coffee. As ICA ceased to operate the industry has become buyer-driven with the presence of players in the coffee chain such as global traders, intermediaries, and retailers of which they emerge as a powerful representative of the supply chain. Gresser and Tickell (2002) ironed out the emphasis of the roasting industry as one of the results of liberalization where five big company roasters lead the industry with 45% of the global operation of the roasted and instant coffee market.

The structural change from the emphasis of producers in the chain to the consuming countries affected the income distribution in the coffee chain. From the 51% of proportion belong to the consuming countries coffee growers earn 20% of the overall coffee revenue to 78% against 13% in favor of consuming countries (Talbot 1997). This situation is attributed to the fact that since the elimination of ICA policies and regulations in the trade of coffee has loosened resulting in increased production globally without taking into consideration the balance between production and consumption which further turns into the oversupply of coffee or the coffee crisis. Hence, coffee farmers struggle with profitability that is worsen by price unpredictability. In the absence therefore of the ICA, the loss of enforcement mechanism created issues on sustainability and challenges concerning how the public goods will be protected that requires collaboration with several sectors to institute changes that bring coffee product the first choice if agro commodity goods.

2.2 *Philippines and the Coffee Industry*

The Coffee business has been operating throughout the country since 1740 and pioneered by a Spanish Franciscan monk in Lipa, Batangas and it has spread quickly within the neighboring places. Subsequently, in 1860 coffee business was propelled in Batangas and extend to catering overseas with European and US markets while locally expanded to Cavite in 1870 having Amadeo town as the first place for the production. The country experienced a market niche in 1880 with the achievement of being the fourth largest exporter of coffee and the main source of beans while Africa and Brazil struggle with rush hit. However; the award was short-lived as the Philippines too was stricken with insect infestation in 1889. It was only in the 1950s that the coffee industry in the Philippines recovered with the advent of more resistant coffee varieties initiated by Americans. The development of the first instant coffee was introduced in the business that resulted in the soaring high of demand especially the coffee beans. This also stabilized the demand for coffee in the 1960s which fuels the farmers to intensify farming activities. But challenges again arise with the global production of coffee resulting in excessive supply from the global market and the country being threatened with imports and competition among producers locally and internationally. The Philippine government has taken action to ban importation to allow local producers to penetrate the local market.

According to the data in 2013 provided by the International Coffee Organization (ICO), about 145 million bags (60 kg/bag) of coffee were produced at US$ 21 billion from countries like Brazil accounting for 34% of the total production and followed by Vietnam with 18%, and Indonesia with 8% respectively (International Coffee Organization 2021). In line with this, the production of coffee in the Philippines accounted for one percent. The Philippines is one of the 60 countries which is located in the so-called equatorial zone "The Bean Belt" and thus, it is considered coffee producers in the limelight although the country is mostly composed of small-scale producers. The statistics from the Bureau of Agricultural Statistics in 2012 revealed that the Philippines produced coffee with roughly 83.5 million trees situated in 119,999 hectares of arable land and the volume of production occupying 64% of the total land area is located in Mindanao. In terms of the number of fruit-bearing trees, 69% of the total has been generated from Mindanao. Specifically, topping the farm area of the coffee farm is SOCCKSARGEN considered as the largest area with 25,223 hectares and seconded by Davao Region with 25,166 hectares, then the ARMM with 13,746 hectares, CALABARZON, 13,563 hectares; and Northern Mindanao with 11,837 hectares (Philippine Coffee Board 2021).

Production-wise, in 2008–2012 has shown an average of 93,173 MT of dried berries output of which the highest contribution comes from Robusta with 71.7% while Arabic has 20.7%. The increase in entire production in 2012 compared to 2011 was 0.47% however; small increase when to compensate for the 6% decrease in 2010–2011 production. It was also in 2012 wherein the SOCCKSARGEN occupied the top spot in coffee production with 31% and seconded by Davao Region (21%); ARMM (12%); CALABARZON (10%); and lastly, the Western Visayas with 6% (Philippine Coffee Board 2021). The Philippines proved to be premier in coffee production and the main exporter in Asia but experienced a decline in 1997 because of the conversion of coffee plantations into other crops and since this year importations began. President Duterte responded to this issue by signing the Philippine Coffee Industry Roadmap 2017–2022 last March 7, 2017, to patronize the country's coffee production for the next five years thus, fully supporting the producers, farmers, and traders. The Industry Roadmap aims to ensure that the coffee industry follows the quality standards globally, environment-friendly, reliable, cost-competitive, and gives sustainable benefits to processors, farmers, exporters, and traders while overcoming challenges over food security and poverty (Philippine Coffee Board 2021).

One of the unique features of the Philippines' coffee production is that the country is among just a few countries that commercialize four varieties of coffee namely the Robusta, Arabica, Excelsa, and Liberica also known as Barako. This is attributed to the country's suitable climate and the soil condition covering from lowland to mountain regions and coupled with the Philippine Coffee Roadmap implementation, the year 2022 will mark the increasing production of coffee into 214,626 metric tons (BusinessMirror 2018) and which possibly made the country self-sufficient with 161% sufficiency level far over the recent 41.6%. This perceptible confidence from the different stakeholders will enable coffee growers to work harder to continuously improve coffee production.

2.3 *Inventory Management Practices*

Nowadays, implementing effective inventory management practices has been the priority of manufacturing companies and in the coffee industry, for instance, one of the determining factors of the success is through the supply chain capabilities of the company. So, these capabilities require an effective inventory management system to overcome competition, achieve and maintain competitive advantage, coffee supply improvement, and provide better quality coffee to consumers. However; sustaining effective inventory management practices is relatively a new concept in coffee production. Undeniably, inventory management shows a significant role in the efficiency and competitiveness of the company especially in enhancing profits by focusing on sales generation through different driving forces. Ohaka and Idoniboye

(2010) stressed that inventory management to be effective is to keep sufficient units of inventory to meet the manufacturing goals and at the same time meet customers' demands as needed. On the other hand, failure to maintain inventory management properly is a failure to meet customers' demands leading to dissatisfaction, loss of market share, and customer switching behavior.

Ondari and Muturi (2016) found out the growing demands for firms to implement inventory management systems not only to improve strategy but attain competitiveness. The inventory system is expected to be a powerful tool in the manufacturing processes of growing beans and their by-products. The advancement of technology has paved the way in transforming the coffee industry into environmentally friendly and more opportunities for improving the industry taking into consideration all the aspects of manufacturing operations. Technology also resolves challenges of the complexity of the manufacturing process through effective inventory management that provides central benefits to the firm. Through the years, the production of coffee has changed dramatically due to the increasing cost of inputs while the prices in the international markets are decreasing. Also, the demand for specialty coffee has soared high that demands a sustainable production of coffee, and where the need to emphasize coffee quality is vital in maintaining sustainable inventory management for greater net returns. To be effective in managing inventory, the management should employ effective management of Inventory tools and techniques, efficient planning, and control mechanisms. Through applying effective management practices, the scope of these practices should be comprehensive which consists of rules and policies for the firms to consistently determine the sufficient quantities in stocks and their replenishment, and regulate policies on holding stocks. Ohaka and Idoniboye (2010) postulated that managing inventory should be connected with the production, sale of product that represents a larger magnitude of the production cost.

On one hand, considering that the inventory is a financial investment of resources, the need to install a system that constituted stock levels by using resources productively for profitable outcomes in terms of ordering, storing, and processing which can also be interpreted with processes related to inventory management through stocks retention within the required level, protecting shares against losses, and proper usage and accounting of the stocks. Meeting the customers' expectations on product availability is essential to maximize the net profit and minimize the cost of investment. Furthermore, the inventory system should be able to monitor the progress of the business in meeting the aim of an efficient and effective inventory management system by measuring consumers' satisfaction, availability of stocks, profits, and expenses (Ibrahim and Zailani 2010). Beske et al. (2006) stated that upstream sustainability by applying effective management of inventory as well as the downstream customer satisfaction.

2.4 Sustainable Competitive Advantage

Attaining a competitive advantage and sustaining it is what almost every business wanted to own. Porter (1995) posits that the core framework for strategy formulation is to use the company's strengths to maximize opportunities and negate the threats while converting weaknesses into strengths to achieve sustainable competitive advantage. Competitive advantage is defined as the result of the company's matching a core competency in terms of superior skills or resources to opportunities in the marketplace (Pride and Ferrell 2018). The company's relative standing in the industry shows whether the company's profit is within the industry average or below. Earlier studies asserted the concept of competitive advantage into an approach or method employed by firms that cannot be imitated by competitors (Dierícky and Cool 1989; Barney 1991). An evident example of this is the resource-based company that applies lower cost of production over rivals and has superior product design. Sustaining competitive advantage is also related to continuous production or persisting over time with strategies and tools to neutralize this threat from competitors by maximizing the use of resources that are difficult to copy.

As applied to the coffee industry, firms that earn more profits than rivals are those possessing superior characteristics, for instance, superior skills and resources that are unique and hard to imitate thus, overcoming the competitors. Barney (1991) stressed that competitive advantage is employing a strategy that focuses on value-creation to be implemented to either potential or current rivals. Barney (1991) emphasized that attaining competitive advantage should resist imitation by rivals and continuous employment of unique value-creation methods synchronously with possible rivals not capable of imitation and also called a dynamic continuum. Coyne (1986) on the other hand, defines sustainable competitive advantage as a company-related endeavor in developing construct wherein customers must understand the vitality of differentiating the company's product offers against the rival's offers and these differences can be brought by the strengths in resources that competitors don't possess. Coyne (1986) added that by predicting the actions of competitors over time, it would easier to match the company's resources to the market opportunities. Sustaining this competitive advantage is made possible by addressing the problem or gap between the resources and skills over opportunities in the marketplace.

Coyne (1986) also identified that sustainable competitive advantage is impacted by three forces: target market size, access to resources and customers in higher capacity, and competitors' power restriction. In normal cases, companies can build sustainable competitive advantage if leaders and managers use methods and techniques that cannot be copied easily. With these, the study on coffee houses aspires to become a leader in the industry by implementing a system of inventory management that promotes sustainable competitive advantage. Such advantages as driven by the basic tenets of cost leadership by obtaining access to raw materials, scale economies, and effective inventory management and monitoring to ascertain that firms can work all references of advantage in costs.

3 Research Methodology

3.1 Research Design

This research used qualitative research with the case study as its main method to solicit responses and ideas from professionals specializing in the supply chain and inventory management experts in various homegrown coffee house outlets. These respondents are experts in their respective fields and possess adequate knowledge of the research topic who can provide accurate and reliable feedback on the case at hand. Furthermore, an in-depth interview was conducted focusing on how effective is the employment of the inventory management practices and the ways and means to address the challenges faced by the industries to come up with feasible and workable recommendations leading to sustainable competitive advantage through qualitative methods (Yin 1994). The suitability of the case study method was anchored on the extent of effectiveness of the application of the current inventory management of the homegrown coffee house coupled with the problems faced towards the reality of the business world.

Also, the paper specifically dealt with qualitative case interviews performed face-to-face with among the two groups of participants. The decision to use the qualitative technique of collecting the information is due to the fact the understanding in-depth how the inventory management practices are employed and sustained and taking detailed responses on practically the challenges confronting the coffee business in continuing their operations. Patton (2002) and Sayre (2001) ironed out that qualitative methods are sued when the authors wanted to focus on insights and explanation, and in-depth understanding. About 125 respondents were obtained through depth interviews and focus group discussion to enable to generalize the findings and increased reliability even with the pandemic crisis. Also, semi-structured personal interviews and focus group discussions were undertaken to ensure that the participants' ideas, opinions, thoughts, attitudes are duly considered during the investigation. Openness in the questions was emphasized to allow the interviewees to express themselves freely as possible. Responses were recorded to make sure that the data presented were accurate and were duly transcribed to reflect the responses in detail and that nothing will be missed from the answers of the participants.

3.2 Sampling and Sample Size

Purposive sampling was undertaken to determine the respondents of the study. The sample size reached 125 workers from different branches of the local homegrown coffee house. These workers were coming from the purchasing and operation staff and those working in the supply chain after considering the effectiveness of the current inventory management practices employed in the coffee house and the challenges faced by them. The status of the employees are regular and have worked with the company since three years back. The distribution of the respondents can be traced as follows: twenty-five (25) respondents from the supply chain department and one hundred (100) staff from the operations and purchasing department assigned in different branches. As reiterated, the respondents were knowledgeable and possess expertise in the current inventory management practices and in manhandling the challenges facing the company in discharging its functions.

4 Results and Discussion

4.1 Effectiveness of Inventory Management Practices

Implementing an effective inventory management system is considered an income booster as too many stocks of inventory held is not favorable and much more with less inventory than the ideal inventory requirements for production. In the study of Rajeev (2008) he found out that companies that adopted effective inventory management proved to have profit growth which also improves working capital-level, customer satisfaction, and production. Emphasis on the functions of inventory management practices can be achieved by linking the company's objectives with the inventory requirement. Heizer and Render (2014) explained that one of the main roles of the inventory management system is transforming the general and broad perspective into day-to-day activities with inventory monitoring and strikes a balance between customer service and inventory investment. Based on the participants' feedback, it was revealed that moderate effectiveness was observed concerning inventory management practices by homegrown coffee growers in raw material sourcing, inventory control, and storage. Even though companies have reliable providers of coffee beans, some suppliers provided sub-standard quality of beans that affects the quality of coffee production. Practices followed by homegrown coffee houses centered on supplier evaluation according to the quantity of purchase, reliability, and price.

In other words, emphasis on quality received less attention which had been existing for the past years resulting to the increasing costs and wastage. Even so, management dealt continuously with miscreant suppliers because of the intention to help alter the quality of the lives of small-scale coffee growers and farmers. Consequently, participants saw this commendable mandate but other suppliers/providers revile resulting significant financial deficits on coffee house especially because of producing low-level quality of coffee beans thus, losing the coffee's freshness of taste (from interview undertaken on April, 2008). Accordingly, Stevenson (2007) emphasized that material quality that are used for manufacturing goods increased the ability in satisfying consumers. Hence, the homegrown coffee house needs to assure that the materials used for manufacturing should meet the needs of the customers for their satisfaction. When it comes to the storage space, moderate effective response was also the feedback on this management inventory practice wherein due to the growing demand of coffee products, the space has become smaller as storage for the raw materials used for manufacturing operations. Results from the interview of one of the interviewees appointed to this role exclaimed that "due to the increasing demand of the coffee house's products as compared to other players in the market, the number of inventories to be maintained steadily increased and we need to maintain product quality and service reliability".

Furthermore, one of the interviewees shared that "the sudden increase in sales has caused sudden requirement of a larger space because of the unforeseen rise of units of materials and stocks which was normally maintained by the coffee house at small space storage" (results from focus group discussion and interview last April 2018). Because of this situation, incompetent usage of inventory took place as every inventory shelf has been fully utilized. This problem increase also because of overstocking of the materials with incorrect labels that makes it difficult to find those inventories when needed which led to loss, material spoilage, and pilferage. Gill et al. (2010) argued that any unused materials or inventory is a liability to the operation and contribute to increasing inventory costs and requires large storage space with high maintenance cost. Maintaining efficient and effective procedures to control inventory is another important inventory management practice where the most of the participants considered as moderately effective (based on the focus group and interview in November to December 2018). Controlling inventory is the technique employed by the homegrown coffee house to optimize inventory use and the aim is to yield maximum revenues by spending lowest amount of investment while maintaining the satisfactory level of consumers.

Analyzing the recent procedures on controlling inventory by the coffee house, the company failed to maintain and monitor the desired level of stocks needed to be profitable and utilize resources productively which must be focused on right ordering and storage. Furthermore, the coffee house find it difficult to predict the ideal levels of stocks that should be maintained daily as well as when the need arises. The weak control system, lack of proper measures to account for materials, and safeguards against losses are problems that affects negatively the company's operational and financial outcomes. Responses from the participants declare that "in some circumstances the lack of materials for manufacturing caused the temporary stoppage of operation" (based on the focus group and interview in November 2018). They recognize the fact that inventory management should be carefully reviewed and improve to ensure that right goods should delivered at the right time, right place, and in superb condition and that, the need to institute a control mechanism for the coffee house that ascertain and keep maximum level of inventory to meet the operational requirement and the performance levels.

Moreover, the participants exclaimed that "if the coffee house adopt and maintain an effective inventory management practices, not only it reduces cost of operation and increase sales but also elevate the competitive position and achieve competitive advantage and sustaining them". Too much stocks withheld means additional costs but keeping inventories that are required or maintaining sufficient materials in the right place and the right time would imply profit optimization (according to the interview in November 2018). These participants' responses align with the study of Naliaka and Namusonge (2015) who stressed that managing inventory impacts positively to competitive advantage of manufacturing companies. Also, Li et al. (2006) pointed out that attaining competitive advantage consists of the ability to enable firms to distinguish itself from the rivals and is also a result of critical decision making by the management.

4.2 *Homegrown Coffee House and Its Challenges Faced*

While the coffee industry in the Philippines is growing through many shops and needless to mention that more than 20,000 outlets that have attributed to an estimated amount of 100,000 million or more revenues annually, it has also confronted with challenges due to increasing competition, demand for efficiency and effectiveness in all spheres of operation (Department of Agriculture 2019). The problem encountered in the inventory management practices of the coffee houses negatively affects how they perform which will be the basis for these companies to reassess their current practices and devise room for improvement. Another major challenge is on acquiring raw materials with the right quality as suppliers are those small-scale farmers who don't have mechanisms and tools to control the quality and used to have improper labeling. In trying to reduce this challenge, coffee testing for quality was undertaken for every delivery of the raw materials made and results show that 10% of materials and inventory delivered are classified as low-grade and did not pass the quality testing because of soil and fungi contamination. The homegrown coffee house's coffee beans that represent the large quantity of inventory are also affected by expiry dates. As previously mentioned, the space of the storage is not sufficient due to the increasing demand for stocks that led to wastage, spoilage, and other incidences that affect overall business performance.

Secondly, price instability of coffee beans has become a multi-facet challenge driven by the heavy fluctuation of prices and other factors that made the business susceptible to losses and negative consequences in consonance with the unexpected rise and fall of the prices of the commodity goods. Quick changes in the market situations affected the supply and demand of the goods especially the coffee beans and price imbalance considering that the high prices of coffee beans benefit the farmers but detrimental to the other actors of the coffee supply chain that includes the homegrown coffee house. According to most of the interviewees, "the instability or the fluctuation of the demand and supply affects the homegrown coffee house as it is forced to charge higher prices to ensure the continuous supply of coffee beans". However; the disadvantage of doing this is the difficulty of the company to change prices because of cut-throat competition with the rivals in the industry. The risk involves in deciding to change price is also a challenge because of its impact on profitability which sometimes companies have no choice but to enforce to maintain their dominance in the coffee retail business.

The last challenging factor that the participants have identified as lapses in the existing inventory management practices by homegrown coffee houses particularly in the performing demand forecasting. The practices of deciding based on the rule of thumb by homegrown coffee houses attributed to the uncertainties and risks in production processes instead of having systematic and quality-oriented standard procedures. Thus, a scientific way of predicting future demand is essential for the company to achieve well-informed decisions in terms of planning, production, and scheduling, and ascertaining that material acquisitions are within the planned production. If the inventory is not adequately managed, making financial decisions will be at risk and at the same time costly, so management should have found ways to overcome this shortcoming to sustain its operations. Furthermore, the interviewees assert that adopting an effective forecasting tool is vital in dealing with the voluminous production of homegrown business coffee to maximize profitability. Lastly, to lessen the challenges on overproduction and underproduction, forecasting efficiency should be given due consideration by the homegrown coffee house to replace the traditionally accepted rule of thumb production decisions.

5 Conclusion

This paper clarifies issues on adopting effective inventory management practices in the context of the homegrown coffee house. It also dealt with how to attain sustainable competitive advantage through upholding an inventory management system that will replace conventional production decision-making using the rule of thumb. Based on the results from the interview and focus group discussion, opinions indicated the importance of implementing effective inventory management practices as an urgent and critical decision that should be acted as first priority if organizations wanted to achieve sustainable competitive advantage to ensure viable and feasible company's performance. General findings dictate that the recent inventory management employed by homegrown coffee houses explicitly demonstrates weak and costly decisions. The consensus of interviewees declared that they need to reassess and review the current inventory management practices to determine areas of improvement, as means to overcome challenges of competition and sustain competitive advantage. Also, deficiency of materials or inventory led to increases in costs, mislead decision-making and unfavorable company performance. Challenges are also prevalent in the industry apart from ineffective inventory practices and these include difficulty to acquire raw materials at the right quantity, right place at the right time, price instability of coffee beans and other commodities, and the lack of the company in implementing the right forecasting method or strategy that caused financial losses. In this study, therefore, it can be concluded that how effective is the inventory management practices connotes a favorable organizational performance which then leads to sustainable competitive advantage.

6 Recommendations

The coffee business in the Philippines is a growing business since its inception and it has experienced opportunities and challenges particularly in addressing the restraint between supply and demand. The main challenge experienced by the home-grown coffee house is on adopting effective and feasible inventory management practices that are expected to yield financial returns. Baron et al. (2010) reiterated that managing inventory is one of the main functions of operational production to address the demands of the customer. The main objective of inventory management is to ascertain that materials are used as required and inventories are disposed of quickly to lessen overstocking, increase inventory costs, and losing sales (Gitmann and Zutter 2012). Within this frame, sets of recommendations were offered are a proposal for improvement to the company's inventory management system in the areas of forecasting, quality output production, and production lead times. These recommendations were summarized as stated:

- Equip employees through regular training on inventory management competencies to effectively manage inventories in accordance with standard quality.
- Adopt a quality culture of applying and sustaining effective inventory management practices.
- Conduct periodic reassessment and review of inventory management practices to determine areas for improvement and ensure their relevance to the new trends in the market.
- Perform benchmarking with other outstanding firms in the conduct of an effective inventory management system.

References

Barney JB (1991) Firm resources and sustainable competitive advantage. J Manag 17(1):99–120

Baron R, Binder A, Wasner G (2010) Neuropathic pain: diagnosis, pathophysiological mechanisms, and treatment. Lancet Neurol 9(8):807–819. https://doi.org/10.1016/S1474-4422(10)70143-5

Beske P, Koplin J, Seuring S (2006) The use of environmental and social standards German first-tier suppliers of the Volkswagen AG. Corp Soc Responsib Environ Manag 15(2):63–75

BusinessMirror (2018) DTI sees bright prospects ahead for Philippine coffee industry. http://industry.gov.ph/dti-sees-bright-prospects-ahead-for-philippine-coffee-industry/. Accessed 1 Mar 2021

Coyne KP (1986) Sustainable competitive advantage: what it is, what it is not. Bus Horiz 29:54–61

Department of Agriculture (2019) 2017–2022 philippine coffee industry roadmap. https://www.da.gov.ph/wp-content/uploads/2019/06/Philippine-Coffee-Industry-Roadmap-2017-2022.pdf. Accessed 11 Apr 2021

Diericky I, Cool K (1989) Asset stock accumulation and sustainability of competitive advantage. Manage Sci 35:1504–1511

Gabriele A, Vanzetti D (2005). Long black surviving the coffee crisis. In: Proceedings of the 49th AARES annual conference, Coffs Harbour, New South Wales, 1–29

Gill A, Biger N, Mathur N (2010) The relationship between working capital management and profit ability evidence from the U.S. Bus Econ J 2020:1–9
Gitman LJ, Zutter CJ (2012) Principles of managerial finance, 13th edn. Pearson Prentice Hall, Boston, USA
Gresser C, Tickell S (2002) Mugged poverty in your coffee cup. Oxford Campaign Reports Series, United Kingdom
Heizer J, Render B (2014) Operations management, sustainability and supply chain management, 11th edn. Pearson, UK
Ibrahim HW, Zailani S (2010) A review on the competitiveness of global supply chain in a coffee industry in Indonesia. Int Bus Manag 4(3):105–115
International Coffee Organization (2021). http://www.ico.org/. Accessed 21 Mar 2021
Kaplinsky R (2004) Competitions policy and the global coffee and cocoa value chains. In: Proceedings of the United Nations Conference for Trade and Development, May 24. Institute of Development Studies University of Sussex, University of Brighton, pp 1–31
Li SR, Ragu-Nathan B, Ragu-Nathan TS, Subba Rao S (2006) The impact of supply chain management practices on competitive advantage and organizational performance. Omega 34(2):107–124
Naliaka VW, Namusonge GS (2015) Role of inventory management on competitive advantage among manufacturing firms in Kenya: a case study of Unga group Limited. Int J Acad Res Bus Soc Sci 5(5):87–104
Ohaka J, Idoniboye EA (2010) Inventory control, sustainability and contemporary challenges of industrial management in Nigeria. Int J Manag Sci 3:15–19
Ondari LM, Muturi W (2016) Factors affecting the efficiency of inventory management in organizations in Kenya: A case of firms in Kisii Town. Int J Econ Commerce Manag 4(4):1019–1034
Patton MQ (2002) Qualitative research and evaluation methods, 3rd edn. Sage Publications, Thousand Oaks, CA
Philippine Coffee Board (2021) Our coffee heritage: coffee's rich history in the Philippines. https://philcoffeeboard.com/philippine-coffee/ 24 Mar 2021
Ponte S (2002) The latte revolution? Regulation, markets and consumption in the global coffee chain. World Dev 30(7):1099–1122
Porter ME (1995) The competitive advantage: a resourced based view. Strateg Manag J 14(3):179–189
Pride WM, Ferrell OC (2018) Foundations of marketing, 8th edn. Cengage Learning, USA
Rajeev N (2008) Inventory management in small and medium enterprises: a study of machine tool enterprises in Bangalore. Manag Res News 31(9):659–669
Sayre S (2001) Qualitative methods for market place research. Sage Publications, Thousand Oaks, CA
Stevenson W (2007) Operations management, 7th edn. McGraw Hill, New York
Talbot J (1997) *W*here does your coffee dollar go?: The division of income and surplus along the coffee commodity chain. Stud Comp Int Dev 32(1):56–91
Taylor PL (2005) In the market but not of it: fair trade coffee and forest stewardship council certification as market-based social change. World Dev 33(1):129–147
Yin RK (1994) Case study research: design and methods, 2nd edn. Sage Publications, Thousand Oaks, CA

Exploring the Efficiency of E-Tendering Services in Oman: Administrative Perspectives from Government and Business Users

Khalsa Al-Bishari and Rashil Khalil

Abstract Information Technology (IT) plays an effective role in managing the government and business entities' operations. Most of the Government institutions have adopted modern IT infrastructures and transferred from traditional administration systems to E-Government service systems. This study evaluates the efficiency of e-tendering services, use of digital administrative transaction system and its impact on the performance of Government authorities and business sector organizations in Oman. To achieve the objectives of study, an online questionnaires distributed to public authorities i.e., Oman Tender Board, external stakeholders such as companies and information technology consultants. The findings of the study reveal that e-system in dealing with administrative transactions has positive impact on the performance of the government and private sector organizations. Despite these benefits, there are some challenges linked with e-system. Based on the Planned Behavior Theory (PBT), our research suggests that successful implementation of digital system requires IT professionals, and strategic plans to implement electronic systems effectively. The research concludes that improving e-infrastructure supported by technical resources is essential to prevent barriers that can impede the efficient use of e-administration. In the light of findings, it is recommended to increase the awareness of important e-administration services among the tender companies, government officials and local citizens.

Keywords E-services · Business users · Organizational performance · Tender board · Information technology

K. Al-Bishari
Faculty Al-Zahra MBA(IT) Program, Middle East College, Muscat, Oman

R. Khalil (✉)
Faculty Department of Management Studies, Middle East College, Muscat, Oman
e-mail: khalil@mec.edu.om

C. J. Chiappetta Jabbour and S. A. R. Khan (eds.), *Sustainable Production and Consumption Systems*, Industrial Ecology,
https://doi.org/10.1007/978-981-16-4760-4_9

1 Introduction

In all stages, e-administration needs planning, developing, applying and conducting of processes through skilled and experienced human resources with IT technical background, infrastructure of solid telecommunication and support from top management. Oman government has the vision to convert the nation into a supportable knowledge society through supporting ICT to develop services of government. ICT has now become the key elements that move forward the improvement process. Hence, we have afforded our care and consideration to finding the national strategy to improve the capabilities and skills of citizens in this field with the objective of improving e-administration services (Al-Hadidi 2010).

This research will call upon all government and private sectors to speedily develop their performance and to simplify their services by using digital technology to usher Oman into the constantly developing domains for applying knowledge. Actually, e-administration has been applied in some institutions, recent studies suggest that e-administration system has not implemented to be applied effectively because there are some obstacles such as citizens' fear of using the electronic system. The Sultanate has applied e-government project and it has some contemporary applications like e-administration and e-tendering. These applications are applied by some institutions but still they encounter some challenges.

This study will analyze the case of applying E-Tenders project in Oman Tender Board to evaluate the impact of digitalization of administrative transactions on organizational performance. As this service is as an electronic service, so the research work will focus more on the issues which are related to technology. There are some challenges that stand in the way of implementing this system, such as the differences in the educational level of employees of the institutions, the fear of illegal information penetration, and the preference of some institutions for the traditional way to complete their administrative transactions. Reforming these situations and commencing conversion process will take time. The main problem may be the employee itself, for example, some employees do not know how to use the computer and use applications to work properly. Lack of important needed knowledge about technology and IT projects. Lack of power supply, shortage of IT skilled manpower and existence of traditional style of administration cause e-administration services to face obstacles to deliver the public services in an effective way (Isaac 2013).

Applying e-administration also includes technical and administrative issues such as the service of internet is slow because of the lack of infrastructure in telecommunication companies in the Sultanate. This is an obstacle in motivating stakeholders to complete their operations, and also the absence of electronic connectivity between the institutions and ministries and the lack of implementing the infrastructure of the Internet (Sarrayrih and Sriram 2015). Moreover, some people don't accept the transformation to e-administration because they don't have confidence about, and also the network is a main reason which discourages citizens from using electronic administration transactions. Additionally, some stakeholders have no awareness about the importance of transition to e-administration and how it will facilitate their life.

Applying e-tendering also has some challenges, such as the problem with less transparency in the public procurement is increasing. The criteria that define the winners are not clear enough. There are some obstacles like heavy documentation, inadequate or poor exchange of information, human errors and long cycle of procurement. All these challenges will result in low performance of the procurement practices. Solutions of e-tendering offer a way to manage and develop the procurement in public administrations. However, because e-tendering is still new, it requires to be introduced in the public market by making effective advertisement of this service (Alyahya 2013). So, this study will investigate some of the challenges which are related to e-administration and e-tendering.

2 Literature Review

Globally, the recent transformation of public sector modernization is linked with the arrival of information technology, which goes under the name of e-administration, a concept that blurs the boundaries between public administration, modern technology, and altering administrative methods. It has no one meaning, however there is general agreement that it aims to improve the administrative services using a diversity of electronic means, and to enhance internal productivity (Castelnovo and Simonetta 2007). Even if perhaps the most extensive meaning of e-Administration is related to satisfaction of customer, its managerial vision is more far reaching, and eventually develops from the idea of a central reorganization of the public division. It is expected that the relationship between public administration and civil society will be altered in its fundamentals by e-Government.

Therefore, the technological improvements which led to the simplification of the administrative practices contribute to enhancing the effectiveness and productivity of the administrative work, so it has complicated and enhanced the difficulty of the government in carrying out the functions and linked services to those improvement which need additional effort to finish the work (Agboola 2013). In this sense institutions recognized the significance of electronic work towards the application of which is identified as E-administration that aims to increase the level of performance and efficiency for institution action and minimize the cost of the managerial process techniques.

Thus, the aim of this research is evaluating the impact of digitalization of administrative transactions on Organizational Performance by identifying impacts of E-administrative transaction on organizational performance, challenges and problems associated with use of traditional system and E-System of administrative transactions, benefits of using e-administrative transaction and role of information technology in e-Administrative transactions. Moreover, it develops a theoretical framework for evaluating the impact of E-administration on organizational performance. According to the findings of research, some recommendations will be given to Tender Board for increasing the use of E-administration through the adequate investigation of the impact of conducting E-administration project.

So, this chapter will review literature on this topic and is organized as follows. An overview of E-administration, an overview of E-Tenders, impacts of E-administrative transaction on organizational performance, problems and challenges of traditional and E-system, benefits of using E-administration transactions and developing a theoretical framework for better evaluating the impact of e-administration on organizational performance. Finally, a summary of the literature reviewed for this chapter is offered.

3 Organizational Performance in Oman Tender Board

The following five empirical studies on the organizational performance in different countries indicate the main criteria to evaluate the success of any organization.

Performance in national and multinational companies in Turkey is the key criteria of evaluating the success, could be well defined as quite a wide concept which measures the ability to attain the target of organization successfully (Zehir et al. 2010). Moreover, Zehir et al. (2010) illustrates that performance can be measured with qualitative measures like innovative performance, organizational commitment and job satisfaction. Also, the performance can be measured with quantitative criteria like profitability, sales growth, return of investment and profitability.

Çiçek and Bilal (2011) indicated that the performance in the United States is automation and developing process of performance and aligns employee‘s improvement and the common aims with shared objectives. Management of performance allows organizations to plan for efforts of employee in support of organizational aims and strategic initiatives and to assess performance and outcomes.

Halimi (2015) in his study of the organization performance in Singapore, stated that organization performance is a wider indicator that contain quality, productivity, consistency. Moreover, measures of performance can contain behaviors, results and relative measures, training concepts and education and tools containing development of management and training for leadership to build important skills and attitudes of performance management.

Al Salmi et al. (2016) in his study "E-Government Analysis: Sultanate of Oman Case", states that the performance of organization depends on employees behaviors which can represent a source of sustainable competitive advantage. High participation of human resource management practices were positively linked to organizational performance. They also claimed that the performance of organization can be affected by various factors which can be identified in two main types:

(1) Internal factors: the factors which are related to the management and control factors.
(2) External factors: which refer to factors outside the control of the institution.

Figure1 Factors affecting the organizational performance

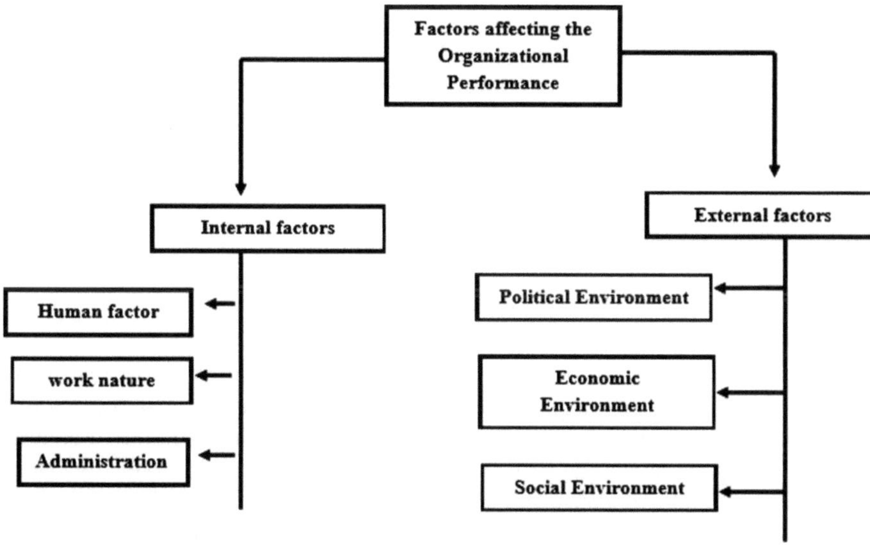

Fig. 1 Factors affecting organizational performance (Kareem 2015)

4 E-Tendering

Many private and government institutions use the traditional paper-based method of tendering. This method has been normal within the institution for an important number of years, but with current technological improvements, this traditional method is rapidly becoming obsolete.

Chilipunde (2013) outlines that the idea of introducing e-tender in some institutions to reduce costs in a sector where the profits are increasing. Moreover, e-tender is becoming a significant tool in the building process of project procurement.

Kerandi (2015) emphasizes that where the process of tender includes the interchange of a huge amount of documentation, created from an electronic format, continuing with paper based tendering is indefensible.

Wagana et al. (2017) points out that in most constructions projects, the documents are replaced electronically. Nonetheless, the constructors are still asking about submitting paper copies of their documents of tender.

According to the "Oman Tender Board", there is necessary need for the tender board of the sultanate of Oman to centralize the online tendering platform for all of its 55 ministries. The objective of the centralizing the platform is to facilitate 24 × 7 availability of access to the tender details and bid submissions, online Pre-bid clarification through clarifications interface, cash less transactions to facilitate all payments in E-Tendering system through the e-Payment gateway of Information Technology Authority under Sultanate of Oman, paper-less e-Tendering management system and post bid online clarifications.

Although most studies supported the use of e-tendering and show the importance of e-tendering, some studies like the study of Wagana et al. (2017) states that still some companies deal with paper copies and prefer to use the traditional system of tendering instead of e-tendering. So, this is one of the challenges of e-tendering.

4.1 What is E-Tendering?

In general, tendering is a process which is undertaken to gain a contract between a client and contractor. E-tendering is well-defined as the practices by which documentation of tender like bills of quantities, drawings and specification are distributed to construction companies in an electronic format and through the internet (Shukla et al. 2016). Thus, the request to tender, award of tender, administration of contract and observing performance of project will be all undertaken electronically. Moreover, queries of tender, updates, assessment of work for payment and notification of payments are all to be replaced electronically through the internet in the system's website and e-mail. Additionally, e-tendering is the practice of issuing documentation of electronic tender to key contractors, sub-contractors and suppliers of industry and getting their responses electronically (Koseva 2012).

E- Tendering Service in Oman (2015) announced that "as part of the Oman initiative, the Government Tender Board has computerized all tendering and procurement processes in the Omani government sector". The main aim of e-tendering is to institute a central state of the art tendering management system and processes. This will help to achieve higher efficiency and will also develop transparency in the process of government tendering.

4.2 Process of E-Tendering:

The solution of E-tendering is conducted through the internet and is predominantly paperless.This practice can contain the advertisement of requirements, production of document, registration of supplier, electronic distribution of documents between buyer and supplier, assessment of submission and finally contract award and publication. Additionally, the solutions of most e-tendering software offers extra support like management of document, archiving, and early warning of chances to suppliers, and maintenance of agreed or potential lists of supplier (Fig. 2).

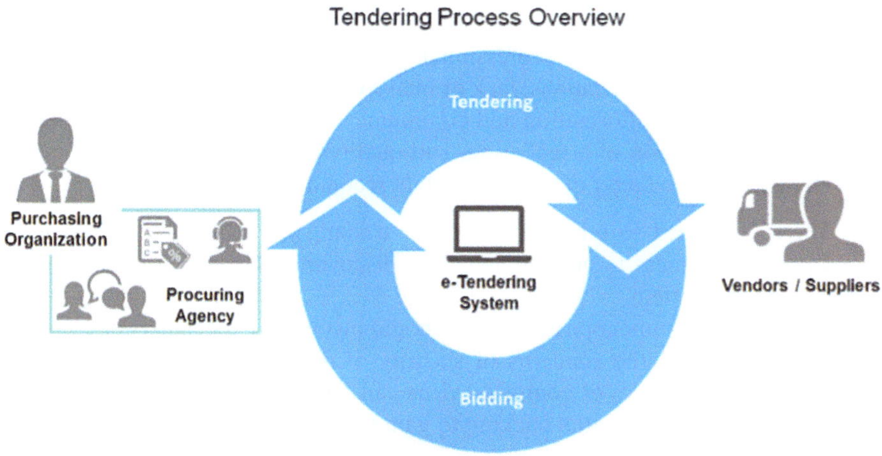

Fig. 2 Overview of tendering process (Saravade 2015)

4.3 *Benefits of E-Tendering:*

The process of e-tendering attempts to remove paper while limiting or restricting the time taken to submit and reply to queries. Interpretations should be delivered electronically and disseminated automatically to all parties, thus decreasing the potential risks and enhancing visibility of replies to all concerned parties. This enhances the accuracy of tenders while decreasing the cost of managing the process of tendering (Tanner et al. 2008).

Nasrun et al. (2016) claim that the benefits of e-tendering include: simplification of process; reducing the period of tendering; fast and precise evaluation, preventing the need for double access of the same information, and decreasing in labor-intensive activities of receipt, decreasing and dissemination of tender documents.

McConnell (2009) states that it is interesting that e-tendering has developed the level of authentication and security of the tender practices. Most solutions of e-tendering software offer extra support like archiving, document management and maintenance of accepted or probable lists of supplier.

4.4 *Barriers of E-Tendering*

Kong and Gray (2006) suggests that one of main issues to deriving the benefits of e-tendering in Malawi is: institutions require to invest money and time to set-up the e-tendering mechanism and to distribute the expected benefits. The presence of internal resistance to accepting a corporate, consistent standard; change management is important to security buy-in; concerns of suppliers through strength of the technology and practical matters like sizes of file; process of a supplier education is important. Upadhyaya et al. (2013) reports that the barriers of e-tendering in Indian manufacturing refer to lack of professionals training, doubt about the system, poor infrastructure of communications, lack of interoperability of software which is used in institution, lack of information and communication technology infrastructure to ensure a secure process of e-tendering, and quality education of graduates is low. This study has offered a set of most important recommendations which are:

1. There is a need to educate and make training for the practitioners in the work environment to embrace recently discovered ICT technologies such as e-tendering system.
2. Building environment professionals should involve the specialists of computer to predict about the concerns of security of delivered sealed bids of tender and frequent attacks of computer virus can be addressed without conducting any questions of potential hacking to gain the confidence of stakeholders in an organization.
3. The stakeholders of any institution should make awareness campaign for the practitioners of administration explaining to them the benefits of e-tendering instead of a system of traditional paper-based tendering to reinforce the need to obtain the technology in their administrative transactions for example e-tendering.

The aim of this study is to identify the importance of resources such as skilled personnel, in order to implement the automation of business transactions. Kerandi (2015) indicates that the challenges or barriers of e-tendering in Kenyan public sector can be related to the integration of software and data management. Software integration belongs to infrastructure of customer's information and its links to suppliers, whereas data management belongs to entry of data coding schema. Moreover, the challenges can be related to system specification such as resources of hardware, resource of network and web server, whereas the challenges in data management are restricted levels of management information about specifications of product and services and expenditure (Laryea and Ibem 2014). It is from this point that technological infrastructure which is made up of hardware and software may affect the use of e-tender in institutions. E-Tender in itself needs some resources like computers, networks, and software. Furthermore management of data is needed as input to the e-tender system (Moe and Päivärinta 2011).

This study attempted to find out whether employees in the institution under the study had the proficiency to make use of the resources for data management in E-tendering. According to Alyahya (2013) the biggest barriers of implementing e-tendering and accepting the use of electronic environment in Saudi Arabia, is the employees themselves. This is because of their lack of awareness of information technology within the construction industry. So, the shortage of skilled employees, and information interchange and cross communication contributed to the challenges of e-tendering (Fig. 3).

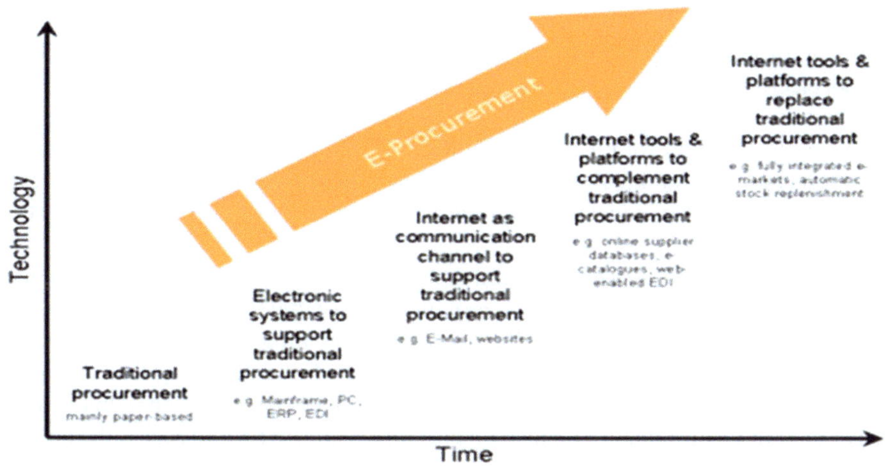

Fig. 3 Barriers of e-tendering (Upadhyaya 2013)

5 E-Administration of Transactions

Karunasena (2012) argues that e-administration in Sri Lanka is considered one of the essential aspects in new societies which was complaining before from inadequate traditional system of administration. E-administration is presented as a new alternative which helps keep pace with current improvement that works to offer relief to stakeholders in various divisions, meet their administration demands and look for achieving their satisfaction.

The study of Onuigbo and Eme (2015) in Nigeria outlined that that orientation to e-administration is not an easy task in many developing countries. This is because there are some factors that can affect to apply e-administration in developing countries. For example, the historic and cultural background of these countries, the infrastructure of IT is poor at organizational and national levels for applying e-administration and lack of real willingness among citizens to use e-administration. So, developing e-administration is a challenge for developing countries.

Riani (2015) states that e-administration in Italy plays an effective and central role in bringing various changes in the administrative work, which contributed to the improvement in many stages of administrative work.

The study of Muñoz-Cañavate and Hípola (2011) in Spain defines e-administration is defined as conducting administrative works by using technology like computer and internet. It can be considered as the administrative labors which contain interchange of information and offering services to people and sector of business with high speed and lower cost over computers and networks with making sure about maintaining the security of information.

Bajramovic (2011) report e-administration initiatives in Bosnia and Herzegovina as it focuses on using a mix of information technology and communication for making all processes of administrative in any organization to develop its performance and improve its competitive position. Additionally, e-administration is determined as applying of technology in all structures of government administration to do all the works electronically.

Nugroho (2015) report Electronic Government in Indonesian reveals that there is an increasing support for positive relationship between information technology and its advantages. He summaries that using IT in administration has a great effect on the performance of organization as it helps to offer a platform for development in many institutions.

Fang (2002) in his paper "E-administration in Digital Era: Concept, Practice, and Development" determined that applying IT in administration of government institutions in Thailand play an effective role. It simplifies the process and tasks of administration by decreasing the use of paper work and replacing the traditional record keeping maintenance to electronic records maintenance which help to easily retrieval of any information of management, staff with less time to access the needed information.

Sarrayrih and Sriram (2015) have discussed that e-administration in Sultanate of Oman is the technological improvement which supports the activities and practices of business and administration. Additionally, offering essential and effective services to the public and stakeholders by using information technologies (like wide area network, internet and mobile computing) which has the capability of transforming the relation with citizens and business.

All the studies state that e-administration is the use of information technology including using internet, network, and electronic record which can facilitate the administration transactions. However, the study of Onuigbo and Eme (2015) in Nigeria addressed that applying e-administration is not a simple task and is considered a challenge for developing countries for some factors like the lack of ICT infrastructure so, they can't implement e-administration effectively. Therefore, the aim of this study to investigate the critical factors which affect implementing e-administration and developing framework which can help developing countries to implement e-development programs that aims to implement an enabling environment for e-administration by developing the capacity of human resources at public organization and developing citizens' ICT readiness.

5.1 Benefits of E-Administration Transaction:

Malipula (2015) concludes in his study that there are five benefits of e-administration: cheaper, more, faster, better and productive. Cheaper: e-administration can produce the similar outputs but at lower aggregate expenses. More: e-administration generates additional outputs at the expenses of aggregate. Faster: e-administration generates the similar outputs with the same aggregate expenses in less time. Better: e-administration generates similar outputs at the same aggregate, but to a better and advanced quality standard. Productive: e-administration generates effective outputs. Ahmad (2013) highlighted that e-administration can introduce the idea of a paperless office, where you can conduct the administrative work at ease by using the e-system, which can save the time, space and be effective.

Renders and Gaeremynck (2012) stated that e-administration enhances the quality and provides satisfactory service to achieve transparency, responsibility, efficiency and timeliness.

Many researchers have claimed that e-administration is basically motivated by the desire to develop the accessibility to services of public, develop efficiency and lower cost. All these studies argued that infrastructures of ICT has great contribution in the role of administration in facilitating the delivery of services in electronic form to citizens in a timely manner and best delivery of service to citizens. Additionally, it facilitates decreasing costs, developing services of citizens and enhancing the efficiency and effectiveness at national and local stages of public sector.

6 Challenges Associated with Use of Traditional System for Administrative Transaction

Ashaye and Irani (2012) identified with two challenges faced by using traditional system (paper-based records). There is limitation of accessibility through hardcopy files because information of paper-based can only be available in a single location at a time. Moreover, he added that only one person has ability to access that information. This serialized approach to information management doesn't work well in today's business environment "I need it now". That is particularly the case since maximum organizations are decentralized and geographically dispersed. When there is only one version of information, dealing with hardcopy documents involved becomes really a physical challenge.

In other hand, Alshehri and Drew (2013) argues that some institutions don't consider traditional system of administration as a challenge because they still prefer the traditional way to complete their administration transactions and they think it is more flexible and more accessible than e-system.

These two studies suggest that using technology for business transactions are more effective. They also report that although traditional system has some obstacles, it also has some advantages like it is simple and can be done by any unskilled person who does not have knowledge of IT.

7 Challenges Associated with Use of E-System for Administrative Transaction

Long (2009) claimed that the lack of IT skilled manpower, lack of power supply and presence of bureaucratic style of governance of administration in Bangladesh are the challenges facing the e-administration. So, this will create a gap between administration and its key stakeholders. The problem behind this is that the gap is increasingly becoming wider to wider and stakeholders lose their trust in the administration. The result of this situation is that the administration is unable to establish effective e-administration.

Therefore, the aim of conducting this research studies to assess the effectiveness of service delivery and to evaluate the impact of e-administration on organizational performance.

Sarrayrih and Sriram (2015) states that there are many organizational, technical, social and financial challenges that are facing e-administration services adoption and dissemination in any private and government institution in Saudi Arabia. There are some studies which support the findings of these studies.

8 Technical Barriers

Isaac (2013) points out that the challenges include poor infrastructure of ICT, security and privacy problems. Some institutions are still lacking in terms of IT infrastructure whereas others require to invest in the installation of modern equipment and require to upgrade the existing infrastructure before considering adopt of the new e-administration. This illustrates that improvement of ICT infrastructure within institutions should be a high priority to implement e-administration. Moreover, Privacy, security and trusting in e-services is another serious mechanical barrier determined in this study.

9 Organizational Barriers

Kerandi 2015 claims that some organizational challenges are associated with the lack of experienced personal and training, Lack of programs to stimulate benefits and advantages of e-administration. Additionally, there exists lack of policy and regulation for e-usage and Resistance to change to electronic methods.

Macueve (2008) defined that promotion is one of the most important factors of successful e-administration systems. When any modern technology appears there should be various steps to motivate and convince people to adopt it and apply it, so the role of institution come in sponsored advertising and promotion which will be an important aid to achieve this task.

The systems of e-administration are modern technological revolutions for various countries around the world and to apply this technology in an efficient way it requires the support of the framework of policy and regulation (Robinson 2015).

Al-Rahbi et al. (2012) says that "To be effective, laws and regulations should cover all applications and related functionality such as e-payments, e-mail usage, copyright rules, e-crimes, e-business, e-commerce and others. The existence and effectiveness of these laws will give all users more confidence and assurance to use e-applications and recommend others to use them.

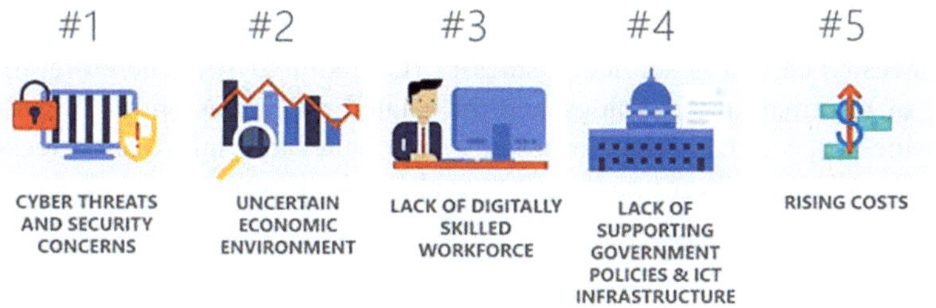

Fig. 4 Barriers of e-administration (Mohammed 2014)

AlShihi (2006) reviews the critical issues of e-administration in Kuwait and reports that the broad scope and complexity of e-administration create many technical and non-technical problems which have to be controlled and monitored. Addressing problems are crucial to the effective implementation of efficient e-administration systems and thus should be examined and planned carefully.

Al-Hadidi (2010) comments in his study that challenges of e-administration in the Sultanate of Oman are due to the lack of sufficient knowledge of the importance of electronic services and its positive impacts in administration transactions. Moreover, the absence of real willingness to provide electronic services and delivery to all the citizens easily also contributes to the situation (Fig. 4).

10 Role of Information Technology

Information Technology investment (IT) is a vital component of any organization performance. If the organization manages IT investment successfully, they will develop performance of organization. Information technology (IT) can be recognized as a distinctive and unique resource in an organization at the end of the twentieth century and 21th century (Abugabah and Sanzogni 2009).

Information Technology (IT) is changing rapidly and its strategic impact on the operations of organization. Additionally, it has been claimed that to be competitive in today's world, organizations need to provide unique services and improve an advanced strategy that serves modern technologies especially Information Technology. These technologies let organizations identify the valuable impacts of information technology as a factor of high standard performance (Magno and Serafica 2001).

The critical challenges that most new organizations face are efficiency, effectiveness, and high quality performance. In the light of these challenges, Information Technology has become an essential strategic component, which helps in creating competitive advantages, and supports survival of organization. On the other hand, organizations are requested to accept modern technologies tools and devices, which support them in attaining more benefits in the economic orientations. The applications of information technology have become one of the important technologies and have engaged significant roles. It has been stated that IT contributes more than 70% of the invested capital in service businesses (Corradini 2010). Therefore, the rapid growth of information technology and the related reflections on the organizations and business, have encouraged a broadly held belief that information technology is central and vital to organizational survival and improvement (Devaraj and Kohli 2003).

11 E-Archives

Ollikainen (2011) defines e-records as a designated repository which reserves for their long-term protection. The Information is documented in electronic form that needs a digitalized system to access or process.

Additionally, E-archiving is the use of modern methods rather than the traditional methods to store folders and important documents and archiving them with help of computer's memory. This modern method developed with the advances of storing mechanisms, memory capacity and the improvement of software, which can retrieve and save information and documents (Al-Shawabkah and Hameed 2013).

11.1 Benefits of E-Archives

According to Stiles (2014) e-archive reduces the use of office paper, warehouse space; so it is ecological as well as more cost effective. E-archive has been estimated to decrease the costs of storage by 25 to 50 per cent and practically decrease retrieval costs.

The study of Manikas (2015) found that e-archive has the feature of electronic storage which allows for more information to be stored in less space with access usually irrespective of time and place. This will improve the access to information, which may lead to an increased organizational memory resulting in higher efficiency and productivity especially in paper-intensive industries.

The benefits of electronic archive can be reviewed by looking at the functionalities offered by applications of e-archive. These functionalities usually vary according to the system used, but the abilities of storing, adding and retrieving documents are considered to be the most essential and fundamental ones (Tagbotor et al. 2015).

Freda (2014) points out that e-archive has the ability to transfer retrieved information to another application for additional processing which can be considered as a basic functionality. Electronic archive also facilitates distribution of information to various geographical locations and to various stakeholders like clients and auditors. For this purpose e-archive are nowadays widely used.

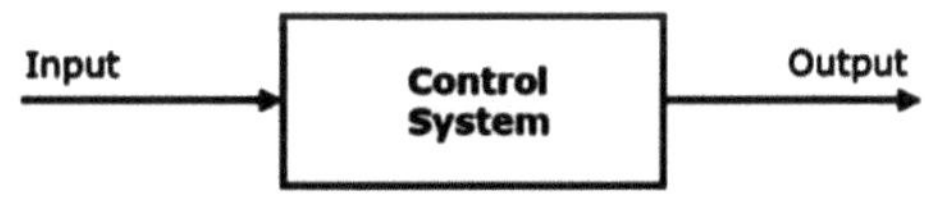

Fig. 5 An overview of control system (Ling 2013)

12 An Overview of Control System

In most organizations, there are devices and control systems. It is essential for the agencies to establish their practices over adopting computer software, which is designed for regulatory activities. This will enhance the ability of the management to alter its strategic plans due to the constant modifications in outside environment. It also helps the managers in their valuation of development made by the organization to achieve its aims (Ling 2005).

A control system is a system, which can provide the wanted response through controlling the output. This figure indicates the simple block diagram of a control system. The control system is represented by a single block and the output is controlled by varying the input (Fig. 5).

13 An Overview of E-Human Resource

Internet has completely changed our social lives and has had an effect on how organizations are managed. It has changed practices of human resource and altered strategies for attracting employees. The strategies have become mainly essential because organizations progressively depend on knowledge, skills, and abilities of workers. Internet has allowed organizations to more cooperative, linked, and responsive to the changing requirements of the workforce (Saigal 2008).

E-HR is the integration of information technologies and services of human resource to automate and facilitate delivery of service (Ma and Ye 2015). Also, E-HRM is a method of establishing strategies of HR, policies, and practices in organizations over an aware and directed support of and with the full use of web based technology (De Alwis 2010).

14 Methodology

This research was motivated by the lack of assessment of technical factors which can affect the service of assigning E-Tenders at Oman Tender Board and the need for developing a modern framework to enhance the performance by addressing the limitations of available frameworks. This framework can guide to preparation of the research goals and research questions for achieving the goals of the research (Fig. 6).

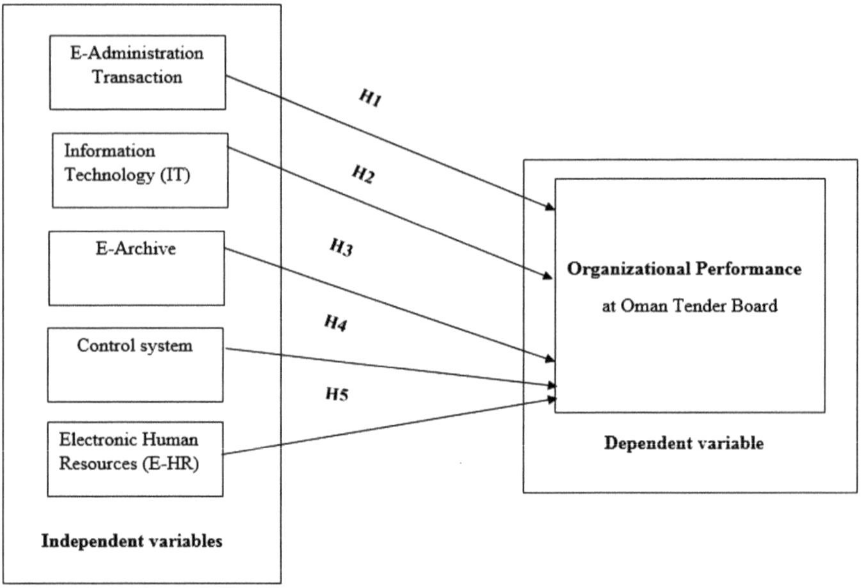

Fig. 6 The theoretical framework for evaluating the performance of e-administration transaction

15 Overview of Hypothesis

As identified in the literature review on the conceptual framework, theoretical framework and observed studies by different researchers, the following hypothesis of research are framed to investigate the impact of e-administration transactions on organizational performance (Table 1).

Table 1 Hypothesis description

Hypothesis	Description
H1	There is a significant relationship between E-administrative of transaction and organizational performance of OTB
H2	There is a significant relationship between information technology and performance of OTB
H3	There is relationship between electronic archive and organizational performance of OTB
H4	There is a significant relationship between Control system and variables of organizational performance of OTB
H5	There is an impact of - human resources and organizational performance of OTB

Table 2 Summary of descriptive statistics
DESCRIPTIVES
Valid cases = 100; cases with missing value(s) = 0

Variable	N	Mean	Std Dev	Minimum	Maximum
P3Q12	100	4.32	0.68	2.00	5.00
P3Q13	100	4.30	0.66	2.00	5.00
P3Q14	100	4.14	0.75	2.00	5.00
P3Q15	100	3.93	0.83	2.00	5.00
P3Q16	100	3.95	0.74	2.00	5.00
P3Q17	100	4.19	0.69	2.00	5.00
P3Q18	100	4.18	0.64	2.00	5.00
P3Q19	100	4.13	0.71	2.00	5.00
P3Q20	100	4.12	0.70	2.00	5.00
P3Q21	100	4.07	0.77	2.00	5.00
P3Q22	100	4.00	0.84	2.00	5.00
P3Q23	100	3.86	0.71	1.00	5.00
P3Q24	100	4.03	0.70	2.00	5.00

Table 3 Summary of descriptive statistics
DESCRIPTIVES
Valid cases = 100; cases with missing value(s) = 0

Variable	N	Mean	Std Dev	Minimum	Maximum
P2Ql	100	3.68	0.95	1.00	5.00
P2Q2	100	3.88	0.81	1.00	5.00
P2Q3	100	3.48	0.96	1.00	5.00
P2Q4	100	3.52	0.97	1.00	5.00
P2Q5	100	3.28	1.03	1.00	5.00
P2Q6	100	3.67	0.85	1.00	5.00
P2Q7	100	3.98	0.97	1.00	5.00
P2Q8	100	4.16	0.91	1.00	5.00
P2Q9	100	3.87	0.81	1.00	5.00
P2Q10	100	3.86	0.85	1.00	5.00
P2Q11	100	3.78	0.73	2.00	5.00

(a) The Relationship Between E-Administration Transaction and Organizational Performance: (H1).

Smith (2003) points out that there is a strong relationship between e-administration transaction and organizational performance. Using technological improvements in

Table 4 Summary of descriptive statistic
DESCRIPTIVES
Valid cases = 100; cases with missing value(s) = 0

Variable	N	Mean	Std Dev	Minimum	Maximum
P5Q37	100	3.94	0.71	2.00	5.00
P5Q38	100	4.05	0.76	1.00	5.00
P5Q39	100	3.89	0.92	1.00	5.00
P5Q40	100	4.05	0.72	2.00	5.00
P5Q41	100	3.93	0.70	2.00	5.00
P5Q42	100	4.17	0.73	2.00	5.00
P5Q43	100	4.05	0.83	2.00	5.00
P5Q44	100	4.06	0.71	1.00	5.00
P5Q45	100	4.15	0.67	3.00	5.00
P5Q46	100	3.77	0.83	2.00	5.00

Table 5 Summary of descriptive statistics
DESCRIPTIVES
Valid cases =100; cases with missing value(s) = 0

Variable	N	Mean	Std Dev	Minimum	Maximum
P4Q25	100	3.74	0.87	1.00	5.00
P4Q26	100	3.64	1.08	1.00	5.00
P4Q27	100	3.78	0.86	2.00	5.00
P4Q28	100	3.37	1.10	1.00	5.00
P4Q29	100	3.60	0.93	1.00	5.00
P4Q30	100	3.89	0.85	1.00	5.00
P4Q31	100	3.82	0.95	1.00	5.00
P4Q32	100	3.86	0.86	2.00	5.00
P4Q33	100	3.71	0.90	1.00	5.00
P4Q34	100	3.63	0.99	1.00	5.00
P4Q35	100	3.53	0.87	1.00	5.00
P4Q36	100	3.72	0.84	2.00	5.00

administration transaction can contribute to the simplification of administrative practices that leads to enhancement of the efficiency and effectiveness of organizational performance.

On the other hand, some studies addressed the issue of e-government and the application of electronic administration in public and private institutions which can affect the performance of an organization. For example, Alshehri and Drew (2013) determine the effect of security threats from both internal and external sources on security of information and its direct and indirect results in the light of conducting e-administration in organizations linked with the safe network. This study involved a range of ministries of Saudi Arabian, which started applying four e-government projects while the study sample contained 148 employees. The results from the study are as follows:

(1) There is a statistically important impact of the technical threats and malicious software in their direct and indirect results.
(2) The presence of statistically important differences for the trends of respondents towards the effect of security threats to the information security and these are recognized according to some variables like gender, age, educational qualification and title of job.

So, it was recommended to provide more attention to improving networks to prevent breakdowns which can shake security of network and giving attention to the significance of internal threats on the security of information which can help to strength the performance of an organization.

Salam (2013) said e-administration could play an important role in developing the quality of services, which are delivered to its stakeholders as it can attain survival, enhance trust and satisfaction and then produce the competitive success for institutions.

Al-Hadidi (2010) stated that the initiatives of e-administration has a direct effect on the citizens and citizens can drive benefits over direct transactions with administrative services. Therefore, if the citizens have the ability to deal easily with electronic administrations, this will affect the performance of organization positively.

From these studies, it seems that there is a strong relationship between electronic administration variable and organizational performance variable. However, the researcher think that there is not always a strong relationship between these two variables because sometimes there are some factors which can affect the operations of e-administration to work effectively and this can lead to poor performance of organization. Therefore, it is necessary for organization to take care about some factors, which can affect the progress of institutions through using technology in their administration transactions.

(b) The Relationship Between Information Technology and Organizational Performance: (H2).

IT leads to motivate the economic growth and improvement of organization performance. IT permits the organizations to be creative in delivery of service that would improve the performance as well as meet customer satisfaction (Ringim et al., 2012). Salam (2013) claimed that IT is one of the main tools that supports good administration by enhancing transparency, accountability and helping to decrease the cost of administration operations. Sharma & Pokharel (2016) suggests that IT capabilities offer the base of gaining competitive advantage that can enhance the organizational performance. A broad body of IT capabilities literature affirm that IT capabilities are a significant resource for efficient collection and utilization of information.

Overall, all these studies show that the variable of IT capability has an important relationship with the performance of organization. So, these studies illustrate most of the correlations between IT capability and organizational performance.

(c) The Relationship Between E-Archive and Organizational Performance: (H3).

E-archive is crucial to the effective and economic operation of organizations. It serve as organizational memory, the indication of past events and base for future action. When implemented, maintained and disposed of in a systematic and orderly fashion it will help to increase the facility to retrieve records easily as well as increase the performance of the organization by enabling it to keep track of its progress (Nengomasha, 2009).

Lyons (2016) claimed that using e-archive information can facilitate easy retrieval by any authorized users and productivity will be developed and costs minimized due to easier access to records. Additionally, less time will be spent looking for required information.

(d) The Relationship Between Control System and Organizational Performance: (H4).

According to the study of Dineshkumar & Kogulacumar (2013), the performance of an organization control system has an important effect on the quality of management guidance. Likewise, companies which disclose inappropriate control system will have high tendency of suffering errors of management in their operations than those companies which report efficient control system. KINYUA (2016) defined that the efficient control system is an essential driver towards gaining the accuracy and quality. Moreover, effective control system has an important role in the success of organization.

(e) The Relationship Between E-HR and Organizational Performance: (H5).

According to Alrahahle (2014) electronic human resources of any organization can contribute to higher performance and become a source of supportable competitive advantage as well-being valuable and rare. Oliveira et al. (2011) claimed that E-HR has the potential to contain these attributes only if we value extremely skilled labor with high technology experiences which can enhance the performance of an organization. Fındıklı & beyza Bayarçelik, (2015) argues that E-HR has powerful ability to process and deliver results in a speedy way and this will save time and resources by getting work done faster and more accurately. So, this will affect the organizational performance positively.

De Alwis (2010) supports that the influence of IT on HR has a negative impact. The finding has also been discussed by other researchers and experts of industry. While some researchers contend that the use of IT by practitioners of HR illustrates the probability of enhancing the productivity of an organization, others argue that the same technology may "have unintended consequences and cause organizations to focus on narrow performance criteria or arbitrary top-down controls. So, this will make the work and process of HR more inflexible. Despite De Alwis (2010) findings that the influence of IT has a negative impact on organizational performance, many other studies illustrate that there is a positive relationship between E-HR and organizational performance (Table 2).

16 Analysis and Results

Impacts of e-administrative transaction on organizational performance.

The above table shows the summary of descriptive statics of the impacts of e-administrative transaction where analyzed individually. The following is critical analysis of some responses linked with literature:

- The questionnaire explored the impacts of E-administrative transaction on organizational performance. Schaller (2016) stated that e-administration permits institutions with traditional systems to develop, test, and maintain application code in a streamlined way when compared to traditional methods of software development. The questionnaire results confirm this view as 47% agreed and 43% strongly agreed that E-administration can increase the value of the organization by developing the way it operates.
- Abugabah and Sanzogni (2009) pointed that using technology in administrative transactions has many quantifiable technical and business benefits like the development cycles will become shorter and the frequency of deployment will increase which leads to increased productivity of an organization's process. The results of the questionnaire confirm what Abugabah & Sanzogni stated as 51% of participants agreed and 40% strongly agreed that E-administration is a way to facilitate developments in an organization's process.
- On other hand Wu (2001), points that IT sometimes does not have a direct influence on sales and profits. It sometimes cost very much so it does not offer any return in investment in a short time. So, it is really hard to evaluate the impact of IT on organizational performance. However, the results of the questionnaires show that most respondents agree that using IT improves efficiency in administrative transactions on organizational performance. For example, 59% of participants agreed and 15% strongly agreed that E-administration will to introduce all the new means and equipment for excellence in performance (Table 3).

17 Challenges of Traditional System:

The above table shows the summary of the descriptive statics generated using the PSPP program for each of the question. The following is critical analysis of some responses linked with some literature reviews:

- Schaller (2016) states that the methods of the traditional system do not allow for requirements of process stages like construction phase or documentation stage to change easily and fast through the development cycle. The questionnaire results confirm this view as 50% of participants agreed and 21% strongly agreed that in the traditional system the process stages like docummentation and construction have a longer duration than other nontraditional administration method.

- Ashaye and Irani (2012) points out that the traditional system has limitations in terms of retrieval of documents through hardcopy files because paper-based information can only be available in a single location at a time. The results of the questionnaire confirm this view as 46% of respondents agreed that the traditional system does not facilitate easy retrieval of documents and information.
- According to Sarrayrih and Sriram (2015) the traditional system becomes more difficult as the number of transactions increase, and does not terminate transactions on time which leads to poor control over the administrative transactions. This drawback has been identified by the respondents of the questionnaire, as 55% of participants agreed and 12% strongly agreed that in the traditional system the termination of transactions is slow and there is poor control over the progress of work (Table 4).

18 Benefits of Using Information Technology in Administrative Transactions

The above table shows the summary of descriptive statics of the benefits of e-administrative transaction where analyzed individually. The following is a critical analysis of some responses in comparison with the literature in this area:

- The questionnaire explored the benefits of using E-administrative transactions. Schaller (2016), states that the e-system emphasizes collaboration, communication and integration between software devlopers and operations of IT in differenet organizations. The questionnaire results confirm what Schaller sated in his study as more than half of the study participants (54%) agreed and (33%) stongly agreed that E-administrative transactions help to share information, ideas and build one mega database between the government and the private sector.
- Renders and Gaeremynck (2012) states that e-administrative transactions improves the quality and provides satisfactory service to achieve transparency and efficiency. The survey results confirm this view as 49% of respondents agree and 31% strongly agree that e-administrative transactions devlops transparency, accuracy and facilitates transfering of information between institution and citizens.

According to Wu (2001) that early adoption of IT in an organization leads to improvement in performance. So, its adoptions generally have an effective impact on organizational effeciency. The results of the questionnaire confirm this view as 57% of respondents agree and 24% strongly agree that using IT in administration will achieve a significant improvement in the innovation and creativity of new technologies (Table 5).

19 Challenges of E-Archives System

The above table shows the summary of descriptive statics of the challenges of E-system where analyzed individually. The following is a critical analysis of some responses linked with the literature:

- According to Schaller (2016), one of the challenges of transforming a bureaucratic environment to an electronic environment requires to the entire culture of an institution to change in order for electronic environment to work. Changing the culture of work can be very difficult for the administration and employees who have performed the tasks of work in non- electronic environment. The results of the questionnaire demonstrate that the majority of respondents (66%) are of the view that the major challenge encountered for the implementation of the e-system is resistance to change. 11% strongly disagreed that the challenge of E-system is resistance to change to e-system. As per questionnaire results of this area of concern, 12% remained neutral, 2% respondents were in disagreement with the query while 3% strongly disagreed.
- Isaac (2013) states that one of the challenges of E-System includes poor infrastructure of ICT, security and privacy of information problems. The questionnaire respondents also identified that of the challenges of E-system is lack of security and privacy of information. The percentages of results of this questionnaire varies as 41% agree, 11% strongly agreed with this challenge. 21% remained neutral, 20% disagreed and 5% strongly disagreed. In fact, according to result of responses of this question of questionnaire this challenge can't be considered as a big issue in Oman Tender Board. Additionally, the result of interview did not show this as a challenge as the interviewee of this organization said that until now Oman Tender Board has not faced any issue in the security and privacy of information.
- Kerandi 2015 claims that some challenges of E-system are associated with the lack of experienced personnel and training and Lack of programs to stimulate benefits of E-System. The results of the questionnaire align with this view as a large percentage of respondents (49%) agreed and (23%) strongly agreed that one of the challenges of E-System is lack of qualified personal and training programs.

20 Critical Analysis of Role of Information Technology in Internal Control System in Managing E-Administrative Transaction

According to Al-Hadidi (2010) IT plays a major role worldwide in its aim to achieve a better internal control system which is one of the main factors which contributes to organizational growth. The questionnaire results confirm this view. When the participants were asked to choose the role of IT in e-administrative transaction, a high percentage (96%) of respondents chose that IT reduces the cost of internal control and thereby develop productivity.

Agboola (2013) states that Information technology has an effective role in the operation of the organization. Additionally, it helps to develop the transparency and efficiency of the organization and create network links for service delivery across institutions. In addition to that, most participants of questionnaire (92%) confirm this view that IT can provide immediate connectivity and develop transparency, efficiency and accuracy.

21 Critical Analysis of E-Human Resources Used in E-Tendering

Sweisi (2010) states that nowadays internet is very important to activate e-services to complete financial transactions, offering bids of tendering, purchasing tendering and offering detailed information about certain tenders. The results of respondents confirm this view as (93%) of study's participants think that one of the most important currently resources used in E-Tendering is portal content and e-services.

AlShihi (2006) reports that the responsibilities of administrative transactions are changing as the as the improvements in technology affect the workplace. Today's changing workplace requires a professional workforce in using the technology effectively. This view is confirmed by the study's participants as 95% of respondent chose Technical Assistance as an important resource in e-tendering. points out that organizations require a culture of information security over and beyond defenses of technology and best practices to ensure the information security assets in organizations. 100% of participants in this study strongly agree with this view that Information security management is the most important technical resource used in E-tendering at Oman Tender Board.

According to the results of the study's participants, it is clear that a low percentage (35%) of funds is allocated for training programs and seminars at Oman Tender Board. This means that this organization does not schedule adequate training programs for its employees to improve their skills in using the technology efficiently.

22 Analysis of Interview

For the interview two officials from e-tendering departments were chosen:

(a) (Manager of e-tendering department)
(b) (Head of e-tendering department)

The interview contains three open ended questions. This interview was conducted with these two officials simultaneously and their responses were recorded. The following text will represent the questions of interview and their analysis.

The Interviewee were Requested to Comment on the Impacts and Benefits of E-Administrative Transaction on Organizational Performance at Oman Tender Board.

The Manager of E-tendering department stated that the effect of E-administrative transaction is very positive. The advantages are transparency of transactions, accelerated tenders, and accessibility of the system which is open 24 h by 7 days a week so it is possible to buy document at any time. The bidding is also done electronically with all the prices attached. By using the electronic system less paper work is done like as all these jobs are done electronically in the system. The system also does the correspondence between the governmental authorities and the private sector which used to be done in person. The correspondence through email and SMSs is fast and helps to accelerate many transactions between the governmental sector and private sector regarding tenders. Another advantage is data reservation and data security as the tender is accessible to specific authorized individuals so its information will not leak as a result to authorization given to specific individuals like the document maker, document reviewer and document approver. Those are the only authorized persons to see the tender document. Through this, the security of the tender and its data is guaranteed. This creates a competition in the private sector to compete in the governmental sector. Furthermore, the registration is done electronically. A month before the company registration expires, the tenders board sends a notification to the company that the registration is about to be over. In addition to that, electronic transactions create some sort of competition and transparency and the right to participate among companies. It also reduces the possibility of human mistakes like selling a construction tender to a services company. There is a possibility of committing mistakes in the transactions done manually, but the electronic system reduces these mistakes.

The head of e-tendering department pointed out the other advantage of the electronic system: facilitating the tracking of the procedure of the transaction at every stage, the ways of enhancement, where it stops and why it is late. This gives the institution a chance to improve and develop. The system also provides a document that details the way to use the system and the way of registration, this makes it easier for the user to open the system at any time. This system also works as a central database that contains all the governmental projects. In the past the ministries which have projects that cost less than 3 million used to follow its procedure by itself and send a report at the end of every year showing that it has given a tender. The Board is only concerned with tenders that cost more than 3 million rials. Now through using the electronic system all these procedures will be in the system and the Board won't have to wait for the annual report, it can follow everything through the system. Those interested in development and planning can note where are the governmental projects are for the seekers and in which sector. Also through the system it will be possible to view the reports and what the projects are up to. The database will benefit many individuals like researchers, decision makers or those concerned with development and enhancement. The system will not be used only by entities subject to the tender law, but also by other companies that has tender law and other entities that are not subject to tender law that want to use the system.

23 Analysis of Response

According to what manager and head of e-tendering department, the impacts of e-administrative transaction on organizational performance at Oman Tender Board is positive. They seem to be convinced that the impact is positive by comparing how the organization suffered when it followed a traditional system and how the situation of institution improved when digital administrative transactions started being used. Also, they illustrated that e-administrative transactions have many benefits which affect the performance of Oman Tender Board positively.

The interviewee were requested to comment on the main challenges linked the use of traditional system and digital system.

Manager of E-tendering department defined the main challenge faced by the Tenders Board in this project: the electronic system not being fully active with some transactions still done manually. The transaction in the manual system requires the concerned staff to review it and put down his notes and any amendments. This procedure takes time, even the follow up is done manually by the Ministry. For example when the document arrives from the Ministry to the Board, the concerned staff may raise some questions while reading the document; so he contacts the document maker usually by phone or Email and this takes time. Through using the electronic system there is a portal which can be accessed by the concerned staff if he has any questions or there are any updates. It can be viewed through the system, and any question can be put in the system. This portal also allows each staff to work on the e-tendering system. They can access it daily and see any new updates, questions or other things. It will make it easier to communicate with the e-tendering system.

The Head of e-tendering department added that the manual tenders are packages of papers, boxes, offers and maps. The Board hasn't yet fully activated the electronic system. Saving all these copies costs a lot, and need special prepared stores. IT also calls for getting rid of extra copies post selling documents. Retrieving documents costs a lot too, as printing all the documents is a big deal and the staff don't want to take responsibility of extra documents. The ministry will not collect any extra copies and the Board doesn't want to keep them because it costs a lot. With the electronic system there are no hard copies; all documents are in electronic format. Now it has become easy, companies buy the document, pay the money and give their offers through the system. When the ministry presents its offers it is visible in the system so the Board won't have to wait for the driver to take the offer. Now it has become faster and safer to communicate. In the past the copies were at risk of being reached by unauthorized bodies, now through the system the Board can control the authorization and choose who can have access to the document. The electronic system can be controlled in case it has been hacked, unlike the manual system which can't be, as copies can be reached or copied without your knowledge.

24 Analysis of Response

The challenges of traditional system are more than the challenges of electronic system. The traditional system becomes more difficult as the number of documents increase, has lack of security and the process of traditional tenders are slow and costly. So, these are big challenges faced in the traditional system which can affect negatively in the performance of board. No doubt, using e-system also it has some challenges but still they are less compared to the traditional system. One of the challenges of e-system is that through the user of e-system accepts to use the system, they still use the manual way to complete their transactions because they do not have the complete confidence to use technology. Also, the big challenge of e-system is that the principle and thought of management is changed when e-system is used.

The interviewee were requested to comment on the important technical resources that used in e-tendering system at Oman Tenders Board.

The Manager of E-tendering department has stated that the technical resources used in e-tendering are represented as the institution applies the system in G-Cloud for all data and systems in the space provided in Cloud so there is no data center and as an institute it doesn't have the resources to manage a data center. This needs specialists and should be in a proper environment like high ground, high roof, cool, spare battery, which means special specifications that costs a lot. Some institutions do not have the man power to manage data centers and the budget to invest in a data center. So the only choice they have is G-Cloud, it has global companies that can provide this service and the space to allow the institute to save all the data and systems in the space provided. E-tendering system is available here so it is connected to the Authority of Information Technology because the Board system is available in the Authority.

The head of e-tendering department also mentioned that other important resources are the specialized human resources in controlling the database. The system of the Board is amendments, requires many amendments in its timetables and additions. Sometimes some paces are not available so the need database expert is needed.

The Board also needs the system manager and the designer of its pages because the system is changing in terms of shape, fields, add forms, and delete forms which need to be designed.

The Tenders Board also needs the projects managers who are dealing with employers in the Tenders Board.

25 Analysis of Response

To conduct e-tendering project effectively and to be activated totally by all the employees, Oman Tender Board should provide good technical resources. Also, it should offer enough and effective resources to apply e-tendering project professionally. The important resources are technical human resources including system manager, system designer and project manager because all these human capitals can play an important role in the success of an e-tendering project.

26 Findings and Conclusion

Findings of the study.

- Most of the respondents agree that the traditional system of administration has many challenges as it is complex, time consuming and lacks security and will negatively affect the performance of the organization.
- It is obvious from the results of participants‘ study that there are positive impacts of e-administrative transaction on organizational performance.
- Despite the success of conducting E-system in an organization, there are also some challenges in using it such as resistance to change to E-system.
- It is noticeable that there are many benefits of using technology in administrative transactions which can enhance the performance of the organization.
- The Board believe that Information Technology plays a significant and effective role in e-administrative transactions as well as on organizational performance.
- The Board provides efficient technical resources to conduct e-tendering project effectively.
- The Board also suggest to add more technical resources in order to activate e-tendering project professionally like allocating electronic human resources, security, activating cloud computing services and allocating specialist in e-tendering like programmer, system designer and project manager.

27 Contributions

This research study investigates the impacts of digitalization of administrative transactions on the organizations performance and the major contributions are:

- The researcher proposes conceptual and theoretical framework for evaluating the performance of e-administrative transactions.
- The research contributes to the field of e-administration research from both the theoretical and practical perspectives. From the theoretical perspective, the

researcher is developing a theoretical framework to evaluate the performance of e-administration. From the practical perspective, this research study offers a comprehensive evaluation the impact of e-administrative transactions on organizational performance of Oman Tender Board.

- The research contributes to a comprehensive literature review using a qualitative method and its role in e-administration research.
- Comprehensive understanding of the research phenomenon being studied and evaluated effectively.
- The researcher is able to establish the factors, benefits, barriers and risk within the context of implanting e-administration based on literature available in e-admin and its impact on the organization.
- The researcher is able to contribute to the literature of e-administration through identification and classification of factors impacting the implementation of e-administration in Oman Tender Board.
- Based on analysis and findings of literature, interview and questionnaire, the researcher has been able to make recommendations for successful evaluation of the impact of digitalizing the administrative transactions on organizational performance.

28 Recommendations

There are a set of recommendations forwarded in light of the findings of the study as follows;

(1) Increasing the awareness of e-administration concept among the citizens. Trying to use social media like TV, Advertisements, internet, and other methods to spread the importance of electronic system.
(2) As suggested by the analysis of the questionnaire that there are not many training programs and seminars that can help the employees to activate e-tendering project effectively. It is recommended to have training centers for citizens to improve the e-literacy.
(3) According to the analysis of the questionnaire, it is clear that the employees of the organization need to be trained lack of adequate training leads to failure in implementing the system. It is recommended to improve IT skills so that they can apply e-administration especially in e-tendering project.
(4) Improving infrastructure of ICT for the city to prevent any weakness related to the network.
(5) Organization should focus on developing Omani employees instead of hiring foreign experts. So, this it will be less expensive, decreasing the rates of unemployment, and enhancing the trust of citizens that the organization is on the Omani side.
(6) Carry out a study on the future of e-administration features and methods to use them to achieve excellence on organizational performance (Hameed 2013).
(7) Focus more on staff training and support to increase technological awareness.

(8) Focus on encouraging managers to use modern technology tools relating to IT and databases.
(9) Promote e-administration services using social media and spread awareness of importance of activating it to benefit the majority of clients. So, this will reflect positively on the organizational performance.

29 Conclusion

The study was conducted with the aim of investigating the impact of digitalizing the administrative transactions on organizational performance at Oman Tender Board especially e-tendering project. Questionnaire for staff and dealers of Oman Tender Board and interview with officials of e-tendering project were conducted to investigate their views about applying the technology in tendering project. Using technology in administration and tenders was considered to be important for managing administrative transactions and tendering project effectively in the institution. Applying e-system in administration has facilitated the e-administrative transactions and e-tendering project deliverables with less hindrances. Highest emphasis was laid on the need of conducting training programs for the staff to be more skillful in using the technology efficiently and to be able to overcome all the challenges they face when they using e-system. The study also concluded the role of information technology in supporting the operations of organization and delivering public services in a more effective and transparent way. Additionally, this study demonstrates the important technical resources that must be used in e-tendering project.

Acknowledgements This chapter is an extended version of MBA-IT Dissertation submitted to Middle East College Oman, by Ms. Khalsa Al-Bishri, Post-Graduate student, supervised by Dr. Rashid Khalil.

References

Abugabah A, Sanzogni L (2009) Information systems and performance: an analytical approach to understand IS value in business organizations. In: Proceedings of the international conference on information systems. Paris, France
Agboola TC (2013) Effects of information technology on organizational communication: a case study of a selected healthcare organization
Ahmad S (2013) E-governance readiness of Upazila administration: a case of two selected Upazila of Comilla district. BRAC University
Al-Hadidi A (2010) An exploratory study on adoption and diffusion of m-government services in the Sultanate of Oman. Cardiff University
Al-Rahbi Y, Al-Harrasi S, Al-Wahaibi S (2012) Technical factors affecting the adoption of e-government
Al-Shawabkah SHA, Hameed S (2013) Role of E-Government in improving organizational performance in the civil status and passports department of Jordan. Int Inst Sci Technol Educ (IISTE) 3(5)

Al Salmi, M., UUM, U. U. M., Aman, K. D., UUM, U. U. M., & Aman, K. D. (2016). E-government analysis: sultanate of Oman case. Open J SocSci 4(3): 35

Alrahahle ASF (2014) The Impact of E-management and the role of human resource development in improving the performance of the organization. Int J Bus Soc Sci 5(10)

Alshehri M, Drew SJ (2013) E-government principles : implementation, advantages and challenges. Int J Electron Bus 12(1):1–17

AlShihi H (2006) Critical factors in the adoption and diffusion of E-government initiatives in Oman. Victoria University

Alyahya MA (2013) Implementing to the efficiency of public construction contract in Moath Ahmad Alyahya implementing e-tendering to improve the efficiency of public construction contract in

Ashaye OR, Irani Z (2012) Supporting the use of E-government in the implementation of land administrative systems: a conceptual framework. EdMedia+Innovate learning. 19–28

Bajramovic K (2011) Implementing e-government in Bosnia and Herzegovina: practices and challenges for the state government

Castelnovo W, Simonetta M (2007) The evaluation of e-government projects for small local government organisations. Electron J E-GovMent 5(1)

Chilipunde RL (2013) Electronic tendering in the Malawian construction industry: the dilemmas and benefits. J Mod Educ Rev 3(10):791–800

Çiçek I, Bilal Ö (2011) The effect of outsourcing human resource on organizational performance: the role of organizational culture. Int J Bus Manag Stud 3(2):131–144

Corradini F (2010) Quality assessment of digital services in E-government with a case study in an Italian region. In: Electronic services: concepts, methodologies, tools and applications. pp. 1100–1118. IGI Global.

De Alwis AC (2010) The impact of electronic human resource management on the role of human resource managers

Devaraj S, Kohli R (2003) Performance impacts of information technology: is actual usage the missing link? Manage Sci 49(3):273–289

Dineshkumar S, Kogulacumar P (2013) Internal Control System and its impact on the Performance of the Sri Lanka Telecom limited in Jaffna District. Int J Adv Comput Technol 2(6):56–64

Fang Z (2002) E-government in digital era: concept, practice, and development. Int J Comput Internet Manag 10(2):1–22

Fındıklı MA, beyza Bayarçelik, E. (2015) Exploring the outcomes of electronic human resource management (E-HRM)? Procedia Soc Behav Sci 207: 424–431

Halimi HS (2015) The role of information technology organizational design in firms ability to innovate

Isaac W (2013) Performance measurement for the e-government initiatives : a comparative study by Willy C Isaac a dissertation report submitted in partial fulfillment of the requirements for the degree of doctor of philosophy in graduate school of computer and information

Karunasena K (2012) An investigation of the public value of e-government in Sri Lanka. RMIT University, Melbourne, Australia, School of Business Information Technology and Logistics College of Business

Kerandi AM (2015) E-government and the transformation of public administration in developing countries: a case study of the kenya revenue authority. Université d'Ottawa/University of Ottawa

Kinyua JK (2016) Effect of internal control systems on financial performance of companies quoted in the Nairobi securities exchange. Jomo Kenyatta university of agriculture and technology

Kong A, Gray J (2006) Problems with traditional procurement in the Malaysian construction industry-A survey. Proceedings Australasian universities building educators association annual conference 2006:1–21

Koseva II (2012) Method that supports e-tendering vendors to enter the Bulgarian public market

Laryea S, Ibem EO (2014) Barriers and prospects of e-procurement in the South African construction industry

Ling Q (2005) Stability and performance of control systems with limited feedback information. University of Notre Dame

Long Y (2009) The impact of information technology on the HR function transformation. University of Twentes
Ma L, Ye M (2015) The role of electronic human resource management in contemporary human resource management. Open J Soc Sci 3(04):71
Macueve GA (2008) Analyzing challenges and opportunities of the implementation of e-government initiatives for development through the lens of the capability approach: case studies from Mozambique. Math Natural Sci Inform Univ Oslo Norway Tech Rep, Dept
Magno FA, Serafica RB (2001) Information technology for good governance. Vol. 8 Yuchengco center for East Asia, De La Salle University Manila
Manikas K (2015) Records management and electronic records management opportunities and limitations: a case study in Greek companies
McConnell DJ (2009) An analysis into the factors affecting the uptake of applications of e-procurement, within the UK public sector. Loughborough University
Moe CE, Päivärinta T (2011) Challenges in information systems procurement in the Norwegian public sector. International conference on electronic government, 404–417
Mohd MN, Roslan S, Salleh NA, Zulhumadi F, Harun AN (2016) The benefits and challenges of E-procurement implementation: a case study of Malaysian company. Int J Econ Financ Issues 6(S7):329–332
Muñoz- A, Hípola P (2011) Electronic administration in Spain: from its beginnings to the present. Gov Inf Q 28(1):74–90
Nengomasha CT (2009) A study of electronic records management in the Namibian public service in the context of e-government. University of Namibia Windhoek.
Nugroho, R. A. (2015). *Electronic government adoption in developing countries: The case of the Indonesian electronic procurement system.*
Oliveira I, Tinoca L, Pereira A (2011) Computers & Education Online group work patterns : How to promote a successful collaboration. Comput Educ 57(1):1348–1357. https://doi.org/10.1016/j.compedu.2011.01.017
Ollikainen E (2011) Electronic archiving applications and their adoption in Finnish accounting firms available on internet. Svenska handelshögskolan
Onuigbo RA, Eme OI (2015) Electronic Governance & Administration in Nigeria: Prospects & Challenges. Arab J Bus Manag Rev (Oman Chapter) 5(3):18
Rachel George M (2015) Exploring the perceived Obstacles to E-administration by faculty members of Eastern Mediterranean university. Eastern Mediterranean university (EMU)-Doğu Akdeniz Üniversitesi (DAÜ)
Renders A, Gaeremynck A (2012) Corporate governance, principal-principal agency conflicts, and firm value in european listed companies. Corp GovAnce: Int Rev 20(2):125–143. https://doi.org/10.1111/j.1467-8683.2011.00900.x
Riani F (2015) Delay in e-government and digitization of public administration in Italy: EU action in Lazio.
Ringim KJ, Razalli MR, Hasnan N (2012) The relationship between information technology capability and organizational performance of Nigerian banks. Int J Bus Res Dev 4(2)
Robinson M (2015) From old public administration to the new public service: implications for public sector reform in developing countries. UNDP Global centre for public service excellence
Saigal A (2008) A study on the impact of information technology use on nonprofit organizations.
Salam MA (2013) E-governance for good governance through public service delivery: an assessment of district e-service centres in Bangladesh. BRAC University
Sarrayrih MA, Sriram B (2015) Major challenges in developing a successful e-government: A review on the Sultanate of Oman. J King Saud Univ-Comput Inf Sci 27(2):230–235
Schaller AE (2016) DevOps transformation challenges facing large scale legacy systems. Utica College
Sharma G, Pokharel M (2016) The Strategic Role Of Information Technology In eGovernment: A Case of Nepal. Int J EBusiness EGovernment Stud 8(1):19–39

Shukla A, Khan MA, Shah M (2016) literature review of adoption of e procurement practices by construction industries
Smith R (2003) The impact of electronic business on the organisation
Stiles T (2014) A guide to archiving of electronic records. Lulu com
Sweisi NAAO (2010) E-government services: an exploration of the main factors that contribute to successful implementation in Libya. University of portsmouth
Tagbotor DP, Adzido RYN, Agbanu PG (2015) Analysis of records management and organizational performance. Int J Acad Res Account, Financ Manag Sci 5(2):1–16
Tanner C, Wölfle R, Schubert P, Quade M (2008) Current trends and challenges in electronic procurement: an empirical study. Electron Mark 18(1):6–18
Upadhyaya P, Mohanan P, Prasad KM (2013) Barriers to adoption of B2B e-marketplaces: an empirical study of Indian manufacturing MSMEs. Rev Integr Bus Econ Res 2(1):555
Wagana DM, Iravo MA, Nzulwa JD, Kihoro JM (2017) The moderating effects of E-government on the relationship between administrative decentralization and service delivery in county governments in Kenya.
Wu F (2001) Bringing "e" to corporate America: the drivers of e-business adoption and its impact on firm performance. The University of Texas at Austin
Zehir C, Muceldili B, Akyuz B, Celep A (2010) The impact of information technology investments on firm performance in national and multinational companies. J Glob Strat Manag 7(1):143–154

Sustainable Consumption and Recycling Practices

Adeel Shah

Abstract Sustainable consumption and recycling practices are a concept to reduce and reuse materials in production. Moreover, the idea envelops the complete supply chain, which incorporates business organizations and consumers. The population rate is increasing at a phenomenal rate, and U.N. estimates that it will reach 11 billion in 2100 besides the consumption and production is already high. The rate of demand and supply for products requires two times the world resources that worrisome and need immediate attention from governments worldwide. Although having the right mindset to recycle and reuse materials, the developed country lacks to become responsible consumers. Average waste created by people in developed and western countries is higher compared to their counterparts. However developing and underdeveloped countries lack the mindset and culture to recycle. All countries need to synergize for a profound and positive effect on the environment globally. The source to create employment and increase economic growth is material consumption remains; however, this requires decoupling resources from the environment for a positive effect. The following chapter discusses sustainable consumption and recycling practices along the same lines.

Keywords Sustainable consumption · Recycling practices · Sustainability · Responsible production

1 Introduction

Sustainable consumption means using resources like water, energy, and raw material most efficiently. It means that business organizations should be responsible for utilizing all resources smartly without wasting during the production process. The concept of sustainable consumption propagates that all kinds of resources are scarce in quantity or impact the environment if used without better management. Business organizations that are manufacturing use biotic or abiotic material as a factor of

---

A. Shah (✉)
College of Business Management, Institute of Business Management, Karachi, Pakistan
e-mail: adeel.shah@iobm.edu.pk

C. J. Chiappetta Jabbour and S. A. R. Khan (eds.), *Sustainable Production and Consumption Systems*, Industrial Ecology,
https://doi.org/10.1007/978-981-16-4760-4_10

production. Besides both materials as input, they also use water, air, and land to create wealth. The biotic materials used as inputs are mostly agriculture products (plants and animals) and abiotic materials, all types of earth minerals, and fossil fuel. Out of many characteristics of the resources, the most important is the scarcity which can upset the ecosystem. The biotic materials need a balanced environment just like humans to survive and reproduce. But as humans, we would need more than an ecosystem to live our daily lives.

The current human civilization relies entirely on electricity. Homes, schools, manufacturing plants, schools, offices, and hospitals all use electricity to operate smoothly. The way of life will, as we know would not exist if there is no electricity. To produce electricity, we either burn fossil fuels, use hydropower, nuclear plants, geothermal, wind, tidal, and solar energy. The materials and energy combine are we able to produce to satisfy demands to create economic profit. However, the inflated market and irresponsible consumption are terraforming the earth to make life difficult to sustain due to ecological footprint, hazardous chemicals, and pollution. Since industrialization, businesses worldwide were focused on producing and customers to consume while both neglected their impact on the future. The only solution to the aggravating situation is sustainable consumption and complimenting strategy recycling used products to slow down the increasing entropy. If adopted, sustainable consumption and recycling practices give green ideas that can be used as the guiding light to implement in all walks of life.

The United Nations has listed the above issue of consuming earth resources in its 17 sustainable development goals for 2030. Goal 12, which is 'Responsible Consumption and Production,' precisely deals and guides the signatory members to achieve it until 2030 (United Nations 2021). Economic growth is directly related to resources and thus imperative to create sustainability for the present and future. It requires an immediate change in our behavior to produce and consume our resources to impact our ecological footprint. Responsible consumption sets the overall tone for sustainability, especially in western countries. The other significant concern regarding consumption is the rate of the growing population of the world.

The United Nations (2020) estimates 8.5 billion in the year 2030, 9.7 billion in the year 2050, and 11 billion in 2100. The population's logic as a primary concern lies in ecological footprint, where 70% of the freshwater consumed to produce agriculture products to feed the current population (United Nations 2021). Wastages are also an important aspect and are directly related to sustainable consumption. The agri-produce currently produce also waste due to food supply chain, process waste, and consumer waste. According to United Nations (2021) estimates, 1.3 tons of food waste annually, whereas 2 billion people are starving in developing and underdeveloped countries. The U.N. goal under discussion also guides to manage shared resources and disposal of toxic wasted and pollutants. The goal further suggests that business organizations and their product consumers in developing and underdeveloped countries recycle and reuse waste to improve sustainability further. A global effort needs the hour to promote sustainable consumption and recycling practice as current practices are no longer viable. The current consumption practices negatively impact the natural environment and resources (see Table 1).

Table 1 Wastages as per United Nations estimates (United Nations 2020)

Item	Waste	Reason
Food	1.3 billion tons of food waste value to USD one trillion each year	Food Supply Chain, retailer, and consumer waste
Electricity	One and twenty billion USD wasted annually	Inefficient light bulbs
Greenhouse gas	Food sector responsible for 22% of greenhouse gas emission	As the forest converts to farm for producing food
Obesity	2 billion people are obese worldwide	Eating more than required due to irresponsible consumption
Other resources	Resources of 3 planets required	Increase global population estimated to reach 9.6 billion till 2050
Drinking water source	3% water is only drinkable and depleting fast	The consumption rate is higher than the natural replenishment rate

The benefits of sustainable consumption and recycling practices positively affect the environment, helping to do better in less. It contributes to reducing poverty, low carbon emissions, and a greener economy.

2 Resource Efficiency

The U.N. goal on sustainable consumption and production also sheds light on resource efficiency (R.E.) (United Nations, International Resource Panel 2018). Resource efficiency purposes benefit realized from the utilization of a resource. Further, it enlightens a decoupling concept between resource utilization and economic prosperity, which will discuss separately in the chapter. The utility derived from R.E. is in the shape of tangible and intangible products or combinations (goods and services).

3 Sustainable Consumption

The consumption calculates by an economy's material footprint; refers to an aggregate of extracted raw materials globally in the complete supply chain to satisfy customer demand (Desa 2018). The material footprint in the developing countries is growing continuously but, in contrast to developed countries, tiny. People depend on materials to satisfy basic requirements like food, clothing, shelter, and other wants. The growing footprint in the developing world signifies that the living standard is increasing. However, it is also essential to reduce virgin raw material consumption while reusing and recycling old material to minimize the environment's impact. It requires a policy to monitor and simultaneously educate business organizations and the public to adopt responsible consumption.

Sustainable consumption addresses the development of sustainability on both upstream and downstream resources. The issue's importance makes it one of the essential goals out 17 as laid down by signing members to achieve until 2030. Goal 12 is 'Responsible Consumption and Production' guides and educates governments, businesses, and people to achieve the set target (United Nations 2021). Sustainable consumption is a versatile goal that spreads every walk of life of our way of life. It sheds light on the behavior of the person in the large-scale organization. The consumption for individual and organization are different. An individual consumer will use them out in the shape of products and services produced by business organizations. However, the individual behavior multiplied by many start influencing the whole equation of demand and supply. Sustainable consumption also emphasis on the management practices by organizational raw material consumption efficiently to create profit. Organizations and individuals should focus on increasing R.E., as explained earlier. Following are the areas where sustainable consumption practices are essential on all levels to impact socio-economic factors.

3.1 Climate Protection

Environmentalist and climate subject experts have extensively published between direct relation resource consumption and greenhouse gas (Aye and Edoja 2017; Garnett, 2009; Neagu and Teodoru 2019; Yang et al. 2017). The relationship exists due to many reasons, and few are deforestation and production of products. The relationship translates to identify that businesses worldwide have designed and developed innovative technologies to better the R.A., enhancing the impact on climate protection.

3.2 Water

Water is precious for human life; only 3% of water is fresh for drinking purposes (United Nations 2021). The 2.5% frozen water in the shape of glaciers and the remaining 0.5% are available in the ecosystem. The current human practices pollute the open water in rivers and lakes faster than nature's capability to replenish it. Besides the cost of polluting water, the other significant cost is on the logistics to supply it. The cost and increasing demand, which is estimated to be a 1% increase per United Nations (2021) estimates, are worrisome. Water is most demanded in the agriculture industry, consuming 69% of the freshwater resource globally. The remaining 19% and 12% consumes in power generation and households respectively. Many unfortunates in developing and underdeveloped countries do not have the accessibility to safe drinking water or poorly managed. The numbers are staggering and soar to 2.1 billion people globally. The United Nations estimates that 844 million lack freshwater services. R.E. in industries like agriculture and power generation, with the help of greener technologies and systems, provides solutions to reduce fatigues on precious resources like water. The production processes are designed to incorporate sustainability to reduce consumption and increase efficiency with minimum usage.

3.3 Energy

The 21st-century global civilization exists because of energy in the shape of electricity. The conventional electricity production methods are costly on the triple bottom line (planet, profit, and people). The feasible and quick traditional way depends burning of fossil fuel. In contrast, hydropower plants like dams produce electricity at a significantly lower rate, but they are costly and long-time projects to start. Companies in the energy sector use clean sources like solar, wind, tidal wave, and geothermal energy to produce and supply energy, creating adverse impacts on the triple bottom line. However, in developing and underdeveloped countries, the economy still largely relies on fossil fuels for electricity production. Many countries do not have the finances and natural resources to build dams, so most rely on fossil fuels. The dams on rivers are very economical but do create adverse effects on the locality's ecological system.

Energy upstream is not the only concern in sustainable consumption, but downstream also plays a crucial role. As estimated by the United Nations (2021), USD 120 billion can save globally by using energy-efficient light source equipment. The consumption patterns for electricity are also a great apprehension in the same context. Western and developed countries demand more energy and are growing faster both for production and household consumption. The rate of consumption is the primary cause of CO_2 emissions and greenhouse effects. Like water, energy also has a disparity in according to the United Nations estimates 840 million people globally do not have access in the year 2017. No excess of fuel and electricity means that many people cannot cook or use technology to maintain their living standards and have no access to adequate education or health care. R.E.'s concept with technological innovations is making upstream and downstream of energy chain sustainable and greener. The products that are electricity are becoming greener, more innovative, and energy-efficient. Developed and western countries need to do more to balance the equation of demand and supply to cater to demand, which automatically influences supply. Consumption behavior will dictate the energy supply chain, which will enable sustainable consumption in true spirit.

3.4 Food Supply Chain

Among natural resources, agriculture and food supply chain is also a natural resource which needs rethinking. R.E. has a more significant role in managing to achieve zero hunger as set by United Nations 17 S.D.Gs till the year 2030. The relation of food supply chain and agriculture has a direct association with sustainable consumption. The relationship exists as land, water, and other resources are used to produce food; also, the opportunity cost of deforestation is global, which is also a major cause of CO2 increase. Like other natural resources, food upstream and downstream is responsible for ecological and societal impact.

The wastages in the chain are worrisome and need to amend to create sustainability. According to United Nations (2021), food waste is very high, and up to one-third of all food-produced waste amounts to 1.3 billion tons of food. The wastage of food is USD one trillion in value. In upstream, harvesting and transportation are the main reasons for waste creation, while mismanagement and irresponsible behavior cause waste downstream by retailers and consumers. According to United Nations (2021), consumers are excessively using food, and an estimates 38 million children under the age of suffering from obesity in the year 2019. Besides the overuse and wastages of food, the production factors are also a concern in sustainable development. The soil's degradation and fertility, excessive use of water, overfishing, and marine environment degradation decreases the natural resources available to sustain earth ecosystems. The current rate of food consumption causes 22% of greenhouse gas emissions. The only way to prevent the deteriorating situation is to benefit from sustainable consumption practices to better our present and the future for our next generation.

3.5 Product Life Cycle

R.E.'s other benefit is on the product life cycle, which divides a product into four broad stages (see Fig. 1).

The consumption of resources occurs in all four stages of P.L.C. The resource consumption starts from the introduction stage and tends to increase in the growth and maturity stage and slow down in the decline stage. The R.E. enables business enterprises to rethink the resource consumption patterns by inducing recycling and reuse concepts in the product design to realize cost–benefit throughout the P.L.C. of the product. The R.E. should be a priority in short P.L.C., for instance, technology-driven products. Smartphones and desktops production has an impact on the environment as the product requires numerous metals. The companies selling such technological products should place a proper recycling system to reverse logistic consumed products to the original equipment manufacturer (O.E.M.) in the chain. It also refers to close loop systems. All components (metals, plastics, and other materials) of the used products will recycle to be used to make the new product. The constant technological development on the products also improves R.E. in our daily lives. For example,

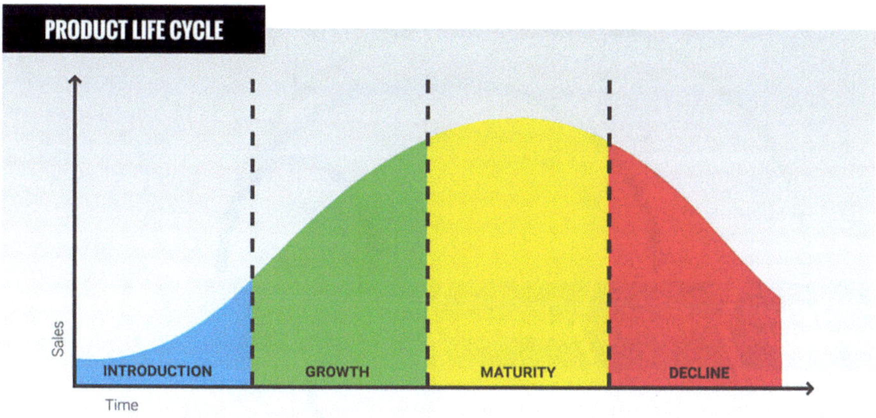

Fig. 1 Product life cycle stages (Sraders 2019)

washing machines for home users use lesser detergent and water to washcloths than earlier technologies. Also, engineers can reduce power consumption by incorporating energy-efficient motor technology (Cunningham 2017; Janeway 2017). The example of the washing machine is relevant for products having longer P.L.C. still increasing R.E. during product utilization and recycling the same will also increase R.E. for manufacturing new products from the raw material.

3.6 Circular Economy

The growing sustainability consumption awareness in consumers and organizations pushes businesses to adopt a circular economy to recycle and reuse their products. The main reasons for the shift are government regulation and societal pressure to reduce materials like plastics single-use in developed countries. However, many developing countries still lack such initiative by policymakers. Thus circular economy is the best model for countries promoting to reduce waste generated during production and reuse the waste material again by recycling (see Fig. 2).

The circular economy model is best suited to complement sustainable consumption as it increases R.E. for the whole product chain (Camilleri, 2018; Kristensen and Mosgaard 2020). The reusing and recycling of materials tend to reduce wastages ending up in landfills. As stated earlier, the concept is still trending; it will take some time in developing countries to adopt a business strategy to reduce overall production and material costs. Companies spend more supply chain cost to source virgin material and transport to the manufacturing facility. The supply chain cost can reduce and give a better contribution margin for the profit, hence improving its triple bottom line. The packaging material used in the product produces with eco-friendly material that either degrades in the natural environment or is recycled to use again in products

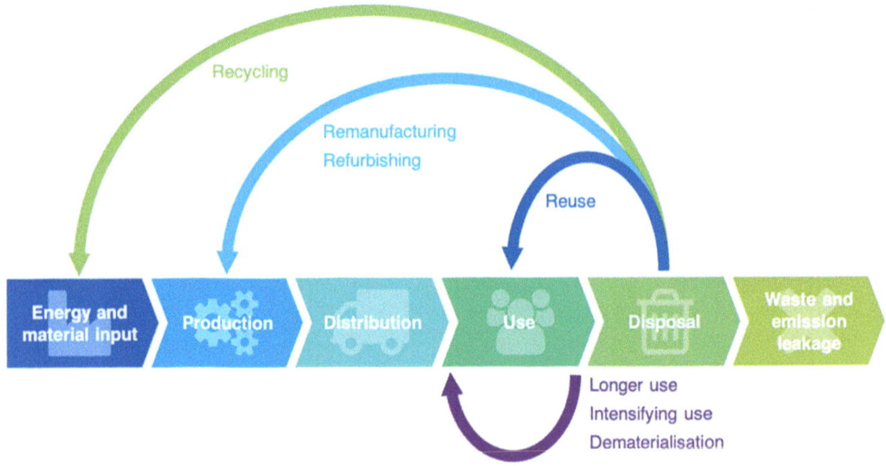

Fig. 2 Reuse from wastage in circular economy model (Geissdoerfer et al. 2020)

as per the circular economy model. Many industries are not upscaling to implement and reap benefits by circular economy model, improving sustainable consumption and meeting 17-SDGs.

3.7 Decoupling

The decoupling strategy is United Nations Environmental Program (UNEP) report (Fletcher and Rammelt 2017). Initially, UNEP reports suggested using fewer materials per unit economic output to reduce the environment's negative impact. However, in the later published report, the definition was made much simpler to focus on lowering natural resources like water and fossil fuel to create wealth and delinked the economic growth from environmental degradation (Panel et al. 2011) (see Fig. 3).

The UNEP, in the context of poverty reduction and elimination, understands the importance of economic growth. Economic growth is essential to employ the world's population, and the rate of people is increasing and will reach 11 billion by the year 2100. Economic growth should equate with the steep rising resource consumption impacting the overall environment and remain the best economic prosperity source. UNEP defines decoupling as an essential precursor to reduce poverty while keeping the resource consumption within planetary boundaries (Hennicke and Hauptstock 2015) and tolerance to avoid any harm to the environment.

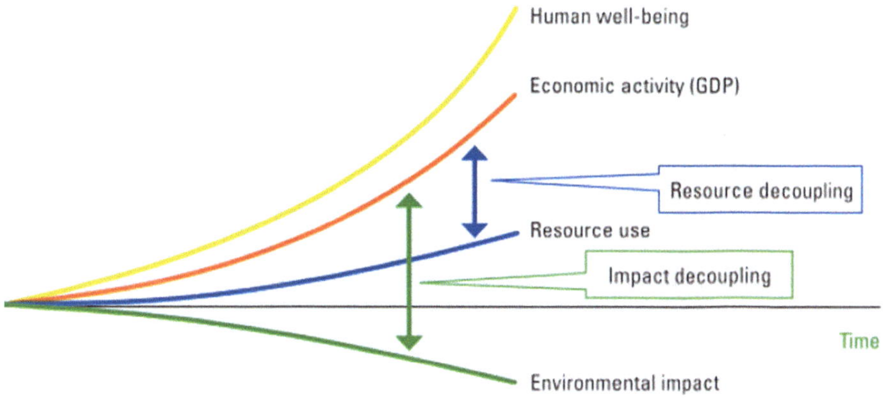

Fig. 3 Impact of resource decoupling on the environment and economic activity (Lutter et al. 2021)

4 Recycling

The process of converting production and consumer waste into raw material, product, and energy is called recycling (Demirbas 2011; Ignatyev et al. 2014; Malinauskaite et al. 2017). The recycling concept is an available alternative to waste disposal to prevent landfills and water pollution, save virgin raw material and reduce greenhouse gas effects (OLEM US EPA 2021). Recycling practices help business organizations, communities, and governments to minimize wastes adding landfills and incinerators greenhouse emissions. It helps to conserve resources like wood, water, and earth minerals. The recycling practices also reduce the economic vulnerability that increases by using a domestic source of virgin materials. Biomaterials are recycled to provide energy production solutions (Paritosh et al. 2017) see (Fig. 4). Recycling has evolved into an industry providing jobs to people (OLEM US EPA 2021).

The recycling practice affects the triple bottom line of many industries. Besides having environmental impacts, as revealed in the earlier paragraph, it also reduces the cost to increase manufacturing contribution margin. The cost factor in manufacturing depends on the recycling efficiency adopted by the organization. However, environmental organizations factor in other factors like the cost of landfills and its fees where it is regulated. Virgin plastic cost less than recycled resin; however, energy consumption is 66% lesser (Plastics Industry Association 2021). Energy consumption is a factor that states recycling efficiency. Recycling materials are done in sequential steps that are continuous loop which is as follows.

Fig. 4 Recycling process

4.1 *Collection and Processing*

The waste material collects several methods, including sidewalk collection, collection centers, and deposit or refund promotions. Business organizations also directly collect recyclables through collection centers and different programs. Once the waste material collection process is complete, the recyclables are sent to the recovery facility for assortment. In the facility, the waste material sorts, clean and processed to convert into a raw material used as manufacturing input. The economic laws of demand and supply work the same way as it affects virgin raw material.

4.2 *Manufacturing*

Many products today are manufactured from recyclables. Most of the everyday household products produced through recycling are paper, paper towels, aluminum, plastics, bottles, and cans. Industries have sophisticated techniques for waste conversion like metals, polymers, rubbers, car tires, and clothing. If the waste cannot convert into raw material, it is sent to use as layers in construction insulations and roads to avoid landfills and incarceration. The plastics in the same cases are converted into park benches or carpeting.

4.3 Products from Recyclables

The recycling loop ends with consumers buying products from recyclables and their consumption. The product packaging has recycled details printed for customer information. The recycling details will either illustrate if the product can recycle or is produce by recyclable.

5 Recycling Practices

Recycling practices are an essential component for sustainability development. It is a modern method to waste reduction. The practice enables recycling all material such as glass paper, metals, plastics, rubbers, textiles, batteries, and electronic equipment. Organic material is also recycled to produce biofuel and fertilizers. The recycling practice enables business organizations to reuse waste to have the same material again. For example, paper recycles to make paper, reducing deforestation for pulp raw material in the paper industry. Metal cans another good example, as companies are reusing cans to produce metal can produce containers without losing any of the required attributes. However, in some cases, not complete waste can convert into raw material. However, some components are salvaged and used in manufacturing either due to high cost, such as lead in batteries.

Industrial waste is a significant concern which in industrialized countries is the biggest waste contributor. To reduce waste accumulation, organizations consume industrial waste to recycle raw material. The industries adopt cost-effective recycling programs. Organizations target components which are commonly used. For instance, for retail stores and warehouses uses cardboards in large quantity for storage and transportation. For such items, companies use recycled cardboard for the purpose. Manufacturers for glass, wood pulp, and paper manufacturers deal directly in common recycle material. There is some manufacturing sector where recycling is low, and they mostly rely on virgin raw material like the metal industry. According to UNEP (2021) report published earlier, metals' use has grown exponentially due to its use in technological devices like mobiles, laptops, cars, batteries, and other everyday use items. Ground metal consumption is high, and the rate is now a threat for the future generation. For recycling metals, the United Nations Environment Program (U.N.E.P) terms metal recycling as urban mining as many metals in used products sit above the ground.

Business organizations are not the recyclers for industrial wastes, but militaries around the world also practice recycling. Old Navy ships are broken down and recycle to use steel from old vessels. The construction industry also uses their industry waste by recycling old concrete to reuse construction to improve profits margin. Fertilizer firms are using biomass to extract energy and convert organic fertilizers to the market. Small Medium Enterprise (S.M.E.) and a municipal corporation in cities collect such waste to supply for processing. Plastics are difficult industrial waste to handle, as it downgrades during recycling. Thus, the reproduce plastic is of low quality; researchers have now developed a process called the viny-loop process that keeps plastic quality the same after recycling.

The consumer waste that is creating pollution is E-waste. The waste is from L.E.D.s, computers, monitors, drives, hard disks, C.D.s, cellular phones; this is a growing problem, especially in developed countries. Much of the toxic E-Waste is dump in the sea. U.S.A. and E.U. now have regulations in place which forbid dumping of E-Waste in sea or land. The E.U. has legislated Electronic Equipment Directive, and the U.S.A. similarly erected National Computer Recycling Act for E-Waste prevention. The recycling practice helps to selvage functional components from devices collected for recycling. Harmful metal like lead is removed from batteries, and the remaining shredded pieces are sent for reprocessing.

6 Conclusion

After announcement of 17-SDGs, business organizations, stakeholders, governments, and nations are becoming more sensitive to the environmental impacts and carbon footprint. The created synergy is to push for greater sustainability. The sustainability created by consumption and recycling practices will impact the present and benefit decades ahead. Businesses implementing sustainable consumption and recycling practices are helping in many ways. The methods and solutions are making supply chains greener, faster, and more customer focus. The techniques discussed in the chapter are to implement in all walks of life. The demand is increasing, and the rate will remain high due to the increasing population. Sustainable consumption and recycling have to be incorporated purposely during the product development stage to reduce virgin raw material usage. Technologies like A.I. and big data will help to adopt eco-friendly production and packaging materials. Many countries in developing and underdeveloped economies lack behind in the process; thus, pollution of land, water, and the environment are high. The developing and underdeveloped countries must erect laws and educate masses for responsible consumption and recycle all materials which are being consumed. The landfills are growing in large number, which is toxic not only to lands but water also. The efforts need to be made part of culture through inculcating the idea at a younger age in children. Governments should rethink and design courses for young children to grow up with the right mindset needed to achieve sustainable development in the world. The right attitude can help countries and societies have forward thought, translating with more profound sustainable business outcomes globally. Making the recycling process more efficient renewable energy sources and using the right technology remains important in the recycling process. The focus is reducing raw material consumption and developing a system that can genuinely provide sustainable consumption and recycling practice in spirit.

References

Aye GC, Edoja PE (2017) Effect of economic growth on CO2 emission in developing countries: Evidence from a dynamic panel threshold model. Cogent Econ Financ 5(1):1379239. https://doi.org/10.1080/23322039.2017.1379239

Camilleri MA (2018) Closing the loop for resource efficiency, sustainable consumption and production: A critical review of the circular economy. Int J Sustain Dev 21(1–4):1–17

Cunningham C (2017) Are energy-efficient washers and dryers worth it. www.Daytondailynews.Com. https://www.daytondailynews.com/business/consumer-advice/come-clean-the-truth-about-energy-efficient-washers-and-dryers/itshnkn6ZClZbZstnFFvOM/

Demirbas A (2011) Waste management, waste resource facilities and waste conversion processes. Energy Convers Manage 52(2):1280–1287

Desa, U. N. (2018). The sustainable development goals report 2018. United Nations.

Fletcher R, Rammelt C (2017) Decoupling: a key fantasy of the post-2015 sustainable development agenda. Globalizations 14(3):450–467. https://doi.org/10.1080/14747731.2016.1263077

Garnett T (2009) Livestock-related greenhouse gas emissions: impacts and options for policy makers. Environ Sci Policy 12(4):491–503. https://doi.org/10.1016/j.envsci.2009.01.006

Geissdoerfer M, Pieroni MPP, Pigosso DCA, Soufani K (2020) Circular business models: a review. J Cleaner Product 123741.

Hennicke P, Hauptstock D (2015) Decoupling resource consumption and economic growth: Insights into an unsolved global challenge. Edward Elgar Publishing, In Handbook of Research on Sustainable Consumption

Ignatyev IA, Thielemans W, Vander Beke B (2014) Recycling of polymers: a review. Chemsuschem 7(6):1579–1593

Janeway K (2017) Yes, your washing machine is using enough water - consumer reports. www.Consumerreports.Or. https://www.consumerreports.org/washing-machines/yes-your-washing-machine-is-using-enough-water/

Kristensen HS, Mosgaard MA (2020) A review of micro level indicators for a circular economy–moving away from the three dimensions of sustainability? J Cleaner Product 243:118531

Lutter S, Giljum S, Gözet B (2021) Decoupling material use and economic performance – materialflows.net. http://www.materialflows.net/. http://www.materialflows.net/decoupling-material-use-and-economic-performance/

Malinauskaite J, Jouhara H, Czajczyńska D, Stanchev P, Katsou E, Rostkowski P, Thorne RJ, Colon J, Ponsá S, Al-Mansour F (2017) Municipal solid waste management and waste-to-energy in the context of a circular economy and energy recycling in Europe. Energy 141:2013–2044

Neagu O, Teodoru MC (2019) The relationship between economic complexity, energy consumption structure and greenhouse gas emission: heterogeneous panel evidence from the E.U. countries. Sustainability 11(2). https://doi.org/10.3390/su11020497

Panel IR, Consumption UNEPS, Branch P (2011) Decoupling natural resource use and environmental impacts from economic growth. UNEP/Earthprint.

Paritosh K, Kushwaha SK, Yadav M, Pareek N, Chawade A, Vivekanand V (2017) Food waste to energy: an overview of sustainable approaches for food waste management and nutrient recycling. BioMed Res Int

Plastics Industry Association (2021) This is plastics: the potential of recycled plastics. Thisisplastics.Com. https://thisisplastics.com/environment/the-potential-of-recycled-plastics/

Sraders A (2019) What is the product life cycle? Stages and examples - TheStreet. Thestreet.Com. https://www.thestreet.com/markets/commodities/product-life-cycle-14882534

UNEP (2021) Resource panel. https://www.resourcepanel.org/. https://www.resourcepanel.org/

United Nations, International Resource Panel (2018) Resource efficiency for sustainable development: key messages for the G20 | green growth knowledge platform. Greengrowthknowledge.Org/. https://www.greengrowthknowledge.org/research/resource-efficiency-sustainable-development-key-messages-g20

United Nations (2020a) Sustainable consumption and production – United Nations sustainable development. www.un.org. https://www.un.org/sustainabledevelopment/sustainable-consumption-production/

United Nations (2020b) Growing at a slower pace, world population is expected to reach 9.7 billion in 2050 and could peak at nearly 11 billion around 2100 | UN DESA | United Nations Department of Economic and Social Affairs. https://www.un.org/development/desa/en/news/population/world-population-prospects-2019.html

United Nations. (2021). Goal 12: responsible consumption and production | UNDP. www.undp.org. https://www.undp.org/content/undp/en/home/sustainable-development-goals/goal-12-responsible-consumption-and-production.html

US EPA, OAR, OAP, CCD (2017) Climate change and the life cycle of stuff. www.epa.gov.

US EPA, OLEM (2021) Recycling basics. www.epa.gov. https://www.epa.gov/recycle/recycling-basics

Yang X, Lou F, Sun M, Wang R, Wang Y (2017) Study of the relationship between greenhouse gas emissions and the economic growth of Russia based on the environmental Kuznets curve. Appl Energy 193:162–173. https://doi.org/10.1016/j.apenergy.2017.02.034

Effect of Sustainable Supply Chain Management on Organization's Performance—Case Study of Logistics and Transport Sector of Pakistan

Fahim ul Amin, Dong Qianli, Jabar Mahmood, and Wasim ul Amin

Abstract This work is aimed to determine the effect of introducing sustainability practices in the supply chain on the performance of organizations in the logistics and transport sector of Pakistan. The logistics and transport sector of Pakistan have a dynamic role in the country's economy and sustainability practices in the sector is believed to be beneficial for the sector as well as society. In this work, state on the art is provided which is followed by the research design. Hypotheses are formulated to determine the effect of sustainability practices on profitability and operational efficiency of the organizations. Regression analysis is performed to validate the hypotheses. Based on the results, some recommendations are also presented in this work

Keywords Supply chain management · Sustainability · Organization's performance · Logistics sector

1 Introduction

The notion of Sustainable Supply Chain Management (SSCM) is gaining popularity in the business community, which is reflected with a substantial growth in the number of companies introducing sustainable practices in their supply chain during the last decade. There are a fairly good amount of theoretical methods and concepts in support of the sustainability requirement in SCM, such as stakeholder theory, transaction cost theory, legitimacy theory, resource approach and agent theory among many. In addition to theoretical justifications, a number of trends in SCM can be highlighted that motivate companies to create sustainable supply chains. These trends include globalization of supply chains, procurement, and production in developing countries, the transition from rivalry among companies to competition among supply chains, an

F. ul Amin (✉) · D. Qianli · J. Mahmood
School of Economics and Management, Chang'an University, Xi'an 710064, China

W. ul Amin
School of Economics and Management, North China Electric Power University, Beijing, China

C. J. Chiappetta Jabbour and S. A. R. Khan (eds.), *Sustainable Production and Consumption Systems*, Industrial Ecology,
https://doi.org/10.1007/978-981-16-4760-4_11

increase in the number of outsourcing operations, limited access to natural resources, natural disasters and macroeconomic risks, to the name few. According to the study integration of sustainability in supply chains helps reduce costs and risks (Ortas et al. 2014). This is primarily due to increased stakeholder loyalty. On the part of consumers, this translates into increased confidence in the company and demand for products. Approval of investors rises the investment attractiveness of the company, and hence the value of its shares, attracting new investors. The state can show its loyalty in the form of tax benefits and other privileges. Thus, a sustainable supply chain helps in reduced operating costs, thereby increasing company net profit.

The logistics and transportation sector of Pakistan have a dynamic role in the country's economy. According to the Bureau of Statistics Pakistan, the logistics sector contributed 12.31% in Pakistan's GDP of year 2019–2020 (Bureau of Statistics 2021) in addition to contributing approximately 7% of total employment of Pakistan. Sustainable practices are reported to introduce in supply chains of logistics sector. With the advent of China Pakistan Economic Corridor (CPEC) project in Pakistan, the sector is expected to grow 3 times in the coming five years. As it is a fast-growing sector that has great impact on the country's economy, it is inevitable to study the effect of sustainability practices on this sector organizations' performance. The organizations in the sector are characterized by the complexity of managing a large network of suppliers and subcontractors, the need for accurate supply, as well as strong dependence on macroeconomic factors and high dependency on technological development. In this regard, it is necessary to determine what changes are being brought by sustainability adaptation on the supply chains of these organizations. This will further require finding out how the implementation of sustainability practices in the supply chain affects the performance of these organizations.

The aim of this work is to investigate the impact of introducing sustainability practices in the supply chain on the performance of organizations in the logistics and transport sector of Pakistan. Firstly, theoretical prerequisites for introducing sustainability practices are presented to get an insight into SSCM. This is followed by a review of similar research in the context of organizations' performance with the aim to determine possible effects of sustainable supply chains. Afterward, indicators of sustainability and organization performance have been determined to taking into account the particulars of the sector. This is an important task since most organizations fulfill only the minimum mandatory requirements of the state and society regarding sustainable supply chains due to the fact that they are not able to determine the contribution of sustainable supply chain practices to the overall result. In the next step, an empirical study has been conducted to determine the effects of implementing sustainable supply chain practices on organizations' performance. The study is performed based on hypotheses developed for this research. Finally, statistical analysis of the research is performed and results are discussed to make practical recommendations for logistics and transport sector. The rest of the paper is structured as follows. Background and state of the art are presented in Sect. 2, where SSCM concepts are presented with significant contribution in literature. The section also discusses the effect of SSCM on organization's performance. Afterward, method is presented in Sect. 3, in which research design, population, scales used in study and research hypotheses formulation is presented. Based on the collected data, statistical analysis is performed and has shown in Sect. 4. Results of the study are discussed in Sect. 5. Section 6 provides the conclusion of the work.

2 Background

3 Sustainable Supply Chain Management

SCM may be defined as the combination of significant business processes from end consumers to suppliers that provide products, services, and information that further create value for the customer and other stakeholders (Croxton et al. 2001). SCM is also defined as a planning and management of all events linked with the search and purchase, management of logistics, coordination, and cooperation with members of the supply chain, which can be suppliers, intermediaries, third-party service providers and by customers (CSCMP 2018). Based on these definitions, it can be summarized that SCM is the integration and coordination of external and internal supply chain elements that create value in order to manage supply and demand within and between companies, as well as to improve the long-term economic results of all participants in the chain.

Sustainability is a development factor that meets current needs without compromising the ability of descendants to meet their needs. Sustainability is defined as the ability to reduce long-term risks associated with resource depletion, fluctuations in energy prices, product liability, environmental pollution (Shrivastava 1995). Sustainability is not limited to ecology only. It also relates to the equality of generations, the priority of the whole over its parts (the balance between the social, environmental and economic aspects of the concept of sustainability), globalization and participation (systematic revision of the sustainability strategy through dialogue with interested parties) (Gończ et al. 2007). Based on the definitions of sustainability, it can be surmised that it is a combination of three components—environmental, social and economic. It is based on a harmonious combination of social economic and environmental goals. Social goals include protecting society, improving working conditions, eliminating discrimination, child labor, that is, those associated with society inside and outside the supply chain. The other important components of sustainability are risk management, transparency, strategy, and organizational culture.

SSCM has been defined in various ways in the literature. Sharfman et al. (1997) have defined sustainable supply chain management as working closely with suppliers to reduce negative environmental effects at all stages of the product life cycle. Carter and Rogers (2008) define this term as a strategic and seamless integration of social, economic and environmental goals in the process of coordinating significant business processes in order to improve the company's economic results and the entire supply chain in the long term. The most comprehensive is the definition proposed by Seuring and Müller (2008), a sustainable supply chain management is the supervision of material, information and cash flows, as well as collaboration of corporations in the supply chain, taking into account the goals of all three aspects sustainable development (economic, environmental and social) and in accordance with the requirements of consumers and other interested parties. A significant amount of literature has suggested that the concept of sustainable supply chains is gaining popularity as an object of study due to the increasing public attention to social and environmental problems. However, studies have not clearly explained the fact that how companies are benefiting from sustainable practices in SCM.

4 SSCM and Organization's Performance

It was a common belief that the main motive for creating sustainable supply chains for most companies is pressure from society or government in the form of minimum sustainability standards. According to a study conducted by Sethi (1979) companies consider sustainability in their activities as a social obligation, responding only to the prevailing legislation and market requirements. Friedman (2007) as an explanation of this fact, indicates the inability of companies to independently determine the number of resources that need to be allocated to social and environmental programs. However, Ortas et al. (2014) on the basis of a literature study revealed that increasing the stability of supply chains, on the contrary, helps reduce costs and risks. Based on this review, it can be concluded that a stable supply chain can serve as a cause of reasonable benefit and contribute to improving the company's financial results if sustainability programs are difficult to reproduce for competitors.

The positive effects of sustainability are obvious, but most companies do not see them. Barnett and Salomon (2012) explained this by examining the relationship between sustainability and financial performance. Companies that are just starting to work towards improving the sustainability of their supply chains have lower financial results than those that do not deal with sustainability at all. Conversely, companies with an already developed system of sustainability programs have the highest results compared to competitors. Based on this dependence, it can be concluded that a positive relationship between a sustainable supply chain and financial results can only be observed in the context of a long-term strategy. Most companies want to see the results immediately, and this prevents them from observing the long-term consequences of introducing sustainability adaptation in the supply chain in the form of higher profits, lower costs and risks. In contrast, the more obvious reason may be the lack of clear quantitative indicators to measure the resistance in this regard, the company is not able to assess its contribution to the financial result and correlate it with the costs. If quantitative indicators can be found to assess the stability of the supply chain, then they are specific to a particular sector, country or company, not suggested to use as a standard. A literature review on the relationship between SSCM and the financial performance of companies suggests that it is different in different studies. The difference is due to different regulatory systems, culture, consumer behavior, and company size (Synnestvedt and Schaltegger 2002). This justifies studying the effects of the policies of the implemented practices of sustainable development of supply chains in the organization's performance in the logistics and transport sector of Pakistan.

5 Research Method

The logistics and transport sector of Pakistan is selected for this study because of its rapid growth and contribution to the country's economy. Due to the globalization trend in supply chains, the sector is largely affected by macroeconomic factors and geopolitical changes. Financial crises, fluctuations in oil prices, wars, and terrorism; all these have a strong impact on the situation in this sector. According to a study conducted in 2019, the biggest concerns for the sector are caused by economic crises (56%), fluctuations in oil prices (50%) and interruptions in raw material prices (48%). In addition, technological changes are expected that will require companies to completely reconstruct their business models and supply chains to remain competitive. In addition to macroeconomic risks and high rates of technological development, the complexity of supply chain management in the sector is associated with a large number of suppliers and the need for high precision supplies. Based on a review of the organizations in this sector, it can be concluded that there are a large number of risks in the SCM. These include macroeconomic and geopolitical risks, upcoming technological changes, difficulties in managing a multi-level network of suppliers and the need for accurate deliveries. One way to minimize these risks is to create sustainable SCM.

6 Research Design

This research is quantitative in nature. The effect of SSCM on an organization's performance is studied based on the features of the supply chains in the organizations, SSCM indicators and performance indicators of organizations in the sector. For SSCM indicators we have used ecological (rational use of resources, emission reduction, use of innovative environmental technologies), social (Workforce, Fundamental human rights, society), and economic factors (Product Responsibility, Corporate governance system, Shareholder Rights, Sustainable Supply Chain Integration). Four indicators of company performance are selected as Return on the Assets (ROA), ROCE (Return On Capital Employed), ROIC (Return On Invested Capital), and inventory turnover.

The regression analysis is performed to determine the most important factors (independent explanatory variables) that affect the value of the dependent variable. SSCM indicators and performance indicators are taken as dependent variables. The size of organizations, profitability, and debt are taken as control indicators.

7 Population

This study is conducted based on the database of the Pakistan International Freight Forwarders Association (PIFFA 2021). There are more than 700 logistics and transportation companies in this database. For this research we have selected 50 companies. In view of the small size of the population, it makes no sense to extract a sample from the data, so firstly the sample and the population were identified for this study. However, given the lack of some financial indicators, as well as indicators of sustainable supply chains for a number of companies in the database, the final sample size is 20 companies. Data was gathered from managers responsible to handle supply chain management.

8 Research Hypotheses

The following hypotheses are formulated in this work:

H1: The results of introducing sustainability practices in the SCM affect the profitability of an organization in logistics and transport sector.

H2: The results of implementing sustainability practices in the SCM affect the operational efficiency of an organization in logistics and transport sector.

9 Analysis and Results

In this work, regression analysis is performed to determine the most important factors (independent explanatory variables) that affect the value of the dependent variable. The regression model is given as:

$$y = \beta_0 + \beta_1 x_1 + \cdots + \beta_{n-1} x_{n-1} + \varepsilon \tag{1}$$

Where y is the dependent variable $x_1, x_2, \ldots, x_{n-1}$ shows independent variables; $\beta_0, \beta_1, \ldots, \beta_{n-1}$ are the regression coefficients, whereas ε shows error, which is factors that are unaccounted for in the model that are random in nature.

10 Descriptive Statistics

Descriptive Statistics of supply chain sustainability indicators are shown in Table 1. It is clear that the highest mean is obtained in the rational use of resources, emission reduction, and use of innovative environmental technologies i.e. a majority of

Table 1 Descriptive statistics of supply chain sustainability indicators

Variable	Mean	Standard deviation	Minima	Maxima
Rational use of resources	57.22409	29.8035	8.527132	99.62121
Emission reduction	62.19499	30.10761	1.893939	98.86364
Use of innovative environmental technologies	61.20687	27.22361	9.469697	99.62121
Workforce	54.93173	30.13538	1.136364	98.86364
Fundamental human rights	54.04008	28.52432	17.05426	97.34848
Society	45.4164	31.6899	0.3787879	99.6124
Product responsibility	55.70432	30.96995	4.545455	99.6124
Corporate governance system	52.83509	27.17612	2.542373	98.51485
Shareholder rights	54.00897	29.59555	0.1074345	98.55769
Sustainable supply chain integration	51.52285	27.33314	6.730769	98.54369

the logistics and transport sector in the implementation of sustainable development practices in supply chains are oriented primarily on the environmental component of sustainable supply chains. However, a large value of the standard deviation contradicts this assertion, since it implies a large scatter of values in the sample. This indicates a high level of uncertainty in the sample, in view of the significant discrepancies within the set compared to the mean value.

Descriptive statistics of performance indicators are shown in Table 2. Productivity ratios of ROA, ROCE and ROIC have arithmetic means of 8.4%, 10, 7% and 16%, respectively. Such high values of performance indicators in the capital–intensive sector indicate the general favorable condition of organizations; however, these indicators for some organizations can be much lower, even negative because of large values of standard deviations. The value of the inventory turnover ratio ranges from 4 to 27, but the mean is 9.5. Taking into account the peculiarities of companies' statistics, this value is quite high.

Table 2 Descriptive statistics of performance indicators

Variable	Mean	Standard deviation	Minima	Maxima
ROA	0.0836506	0.0660177	– 0.0247717	0.3774968
ROCE	0.1065045	0.1043491	– 0.249396	0.3614955
ROIC	0.1597782	0.123995	– 0.09092	0.4501
Inventory turnover	9.511881	5.297298	4.025339	27.38614

11 Regression Analysis

Regression analysis is performed in order to test the formulated hypotheses. Table 3 shows the regression analysis of Hypothesis H1 in which ROA, ROCE, ROIC are used as dependent variables. At the first stage of the regression analysis, base models are constructed as a reference to ensure that the results obtained after adding stability indicators to the base model are not affected by external factors. The first three columns in the Table represent so the results of the regression analysis of base models for ROA, ROCE, and ROIC, respectively, and the last three columns show hypothesis H1 test results.

In the base linear regression model for ROA, the adjusted determination coefficient is 0.48, and all the coefficients in the model are statistically significant. The model for studying the influence of the policies of the implemented practices of sustainable supply chain development on the profitability of companies, built on the basis of the basic model, where ROA is taken as an indicator of performance, explains 64% of the changes in the studied parameter. Among the ten indicators of supply chain sustainability, only two were statistically significant—society and product responsibility.

In the second model examined, ROCE is used as an indicator of the profitability of companies' owners. As in the previous case, in the base model, all ratios are statistically significant and account for 45% of changes in the rate of return on capital employed. When the supply chain stability indicators are added to the model, the adjusted determination coefficient increased to 0.81. Such a high value of the indicator allows us to talk about the high significance of the model. Five out of ten indicators of supply chain sustainability in the model are significant. Based on the results obtained, a positive linear relationship was established between the Rational use of resources, Society, Workforce, and ROCE indicators, and a negative between the emission reduction, Product responsibility, and ROCE indicators.

In the third model, the impact of introducing sustainability adaptation in the supply chain on the profitability of these companies was estimated using the ROIC. In the base model, the adjusted determination coefficient, as in the previous two models considered, is not high 0.53; however, when indicators of supply chain stability are added to the model as independent variables, this coefficient increased to 0.90. Given that the maximum value of the coefficient of determination in an ideal model is 1, the resulting coefficient indicates a very high level of significance of the regression model. In this model, five significant coefficients were found for indicators of supply chain sustainability, among them, positive ones are Workforce, Society, and negative ones are Emission reduction, Use of innovative environmental technologies, and Product responsibility. In short, results show that hypothesis H1 is partially acceptable.

Table 4 shows the regression analysis of Hypothesis H2 in which inventory turnover indicator is selected as an indicator of the performance of organizations in the logistics sector whereas supply chain stability indicators are taken as independent variables.

Table 3 Regression analysis of hypothesis 1

Variables	ROA	ROCE	ROIC	ROA	ROCE	ROIC
Rational use of resources	–	–	–	0.0007843	0.0078815*	0.0021656
Emission reduction	–	–	–	– 0.0007019	– 0.0025018*	– 0.0021515*
Use of innovative environmental technologies	–	–	–	– 0.0003988	0.0006615	– 0.001977*
Workforce	–	–	–	0.000274	0.0015488*	0.0020088*
Fundamental human rights	–	–	–	– 0.0002886	– 0.000305	– 0.00082
Society	–	–	–	0.0007449*	0.001563*	0.0014757*
Product responsibility	–	–	–	– 0.0008399*	– 0.0029513*	– 0.0015009*
Corporate governance system	–	–	–	0.00025	– 0.0000962	– 0.0004654
Shareholder rights	–	–	–	0.0002939	0.0000937	0.0000573
Sustainable supply chain integration	–	–	–	0.000208	– 0.0007515	– 0.0001487
Total assets	– 0.0131224*	– 0.0149914*	– 0.0308327*	– 0.0052515*	– 0.0091547*	– 0.0010804*
Net gross margin	0.2544377*	0.3465868*	0.385197*	0.2695179*	0.1404133*	0.6267201*
Debt/assets	– 0.1548102*	– 0.3632056*	– 0.1649229*	– 0.2460063*	– 0.5726096*	– 0.3287424*
Constant	0.3860073*	0.5240104*	0.8436272*	0.2169712*	0.3515599*	0.2337421*
P-value	0.0000	0.0000	0.0000	0.0000	0.0000	0.0000
R2	0.4933	0.4872	0.5636	0.6933	0.8641	0.9325
Adjusted R2	0.4760	0.4487	0.5308	0.6422	0.8052	0.9032

Table 4 Regression analysis of hypothesis 2

Variables	Inventory turnover	
Rational use of resources	–	– 0.0127907
Emission reduction	–	– 0.0097001
Use of innovative environmental technologies	–	– 0.0396406
Work force	–	– 0.0149594
Fundamental human rights	–	0.0172646
Society	–	0.0886108[a]
Product responsibility	–	0 0.0011495
Corporate governance system	–	0.0345118
Shareholder rights	–	0.0588328[a]
Sustainable supply chain integration	–	– 0.0130867
Total assets	– 0.7286314[a]	– 0.4678751[a]
Net gross margin	– 6.675429[a]	– 10.49935[a]
Debt/assets	– 6.091712[a]	– 11.69303[a]
Constant	29.90427[a]	21.16056[a]
p-value	0.0013	0.0000
R2	0.3591	0.6784
Adjusted R2	0.3311	0.5936

[a] *Statistically significant level of 5%*

In the base model, the coefficient of determination is very low; the independent variables in the model account for only 33% of changes in the dependent variable of inventory turnover. When administered to the model parameters of stability of supply chains adjusted coefficient of determination found at 0.59. Based on this value, we can conclude that the hypothesis H2 is acceptable. Statistically significant coefficients were obtained before the following variables—Society and Shareholder Rights, and in both cases, there is a positive linear dependence of inventory turnover on these explanatory variables.

12 Discussion

According to the results of the regression analysis, only a part of supply chain sustainability indicators has a substantial positive effect on the performance of organizations in the logistics and transport sector. In particular, in order to increase profitability, organizations should focus on the introduction of sustainability practices at all levels of SCM in the following areas:

1. Rational use of materials, water, and energy;
2. Employees satisfaction with work, diversity and equal opportunities for all employees, wage levels, health and safety, employee training and development;
3. Maintaining the organization's reputation, improving living conditions and protecting society, adhering to business ethics.

The negative impact of some indicators on profitability can be explained by the high costs of introducing certain sustainability practices. On the other hand, the reason for such results maybe because of the relationship between indicators of supply chain stability with negative ratios and company performance. In short, organizations that already implementing sustainability practices in supply chains and organizations that do not have such practices have higher profitability than organizations that are just starting out on the sustainability practices in the supply chain.

To confirm this assumption, it is necessary to construct additional regression models with similar indicators, but with a time lag. It will help to confirm or reject the long-term effects of implementing sustainability practices in the supply chain. However, based on the results of the regression analysis, organizations in the logistics and transport sector should pay attention to reducing emissions of pollutants, introducing innovative environmental technologies and processes, and improving quality and product safety. These practices require significant investments, which may pay off only in the long run.

According to the results, it is revealed that organizations' efficiency can be positively influenced by incorporating sustainability in the supply chain in the following areas:

1. Equity of shareholders; restriction of the use of instruments to combat acquisitions
2. Protecting interests and improving the living conditions of society, maintaining the organization's reputation, adhering to business ethics (the absence of bribes and corruption at all levels of the supply chain).

Thus, to incorporate sustainability in the supply chain, organizations should focus on the following areas: workforce, rational use of resources, shareholder rights and society. These practices have loads of positive effects on the profitability, and operational efficiency of organizations in the logistics and transport sector.

13 Conclusion

This work was aimed to investigate the impact of sustainability in the supply chain on the performance of organizations in the logistics and transport sector of Pakistan. In this work, the state on the art was provided which was followed by the research design. Two hypotheses were formulated to determine the effect of sustainability practices on the profitability and operational efficiency of organizations. A regression analysis was performed to validate the hypotheses. Results exhibit that hypothesis

H1 is partially accepted whereas hypothesis H2 is accepted. Based on the results, it is recommended that organizations should focus on the workforce, rational use of resources, shareholder rights and society as it will have a positive effect on profitability, and operational efficiency of organizations. The results of this study can be used as an incentive for the logistics and transport sector of Pakistan to incorporate sustainability in the supply chain. Recommendations based on the results of the study will be useful to organizations as a starting point for auditing practices and increasing the sustainability of supply chains. In addition, the results can serve as the basis for further research on the effects of sustainable supply chains.

References

Barnett ML, Salomon RM (2012) Does it pay to be really good? Addressing the shape of the relationship between social and financial performance. Strat Manag J 33 (11): 304–320. https://doi.org/10.1002/smj.1980

Bureau of Statistics, Pakistan (2021) Sectoral shares in GDP. https://www.pbs.gov.pk/sites/default/files//tables/national-accounts/Table-7.pdf

Carter CR, Rogers DS (2008) A framework of sustainable supply chain management: moving toward new theory. Int J Phys Distrib Logist Manag. https://doi.org/10.1108/09600030810882816

Croxton KL, García-Dastugue SJ, Lambert DM, Rogers DS (2001) The supply chain management processes. Int J Logist Manag 12(2):13–36. https://doi.org/10.1108/09574090110806271

CSCMP (2018) CSCMP supply chain management definitions and glossary. Cscmp 2018. https://doi.org/10.1007/978-1-349-95810-8_371

Friedman M (2007) The social responsibility of business is to increase its profits. In: Corporate ethics and corporate governance, pp 173–78. https://doi.org/10.1007/978-3-540-70818-6_14

Gończ E, Skirke U, Kleizen H, Barber M (2007) Increasing the rate of sustainable change: a call for a redefinition of the concept and the model for its implementation. J Cleaner Product 15(6):525–537. https://doi.org/10.1016/j.jclepro.2006.05.018

Ortas E, Moneva JM, Álvarez I (2014) Sustainable supply chain and company performance: a global examination. Supply Chain Manag 19(3):332–350. https://doi.org/10.1108/SCM-12-2013-0444

PIFFA (2021) Database of the Pakistan International Freight Forwarders Association. https://www.piffapk.com/members.php

Sethi SP (1979) A conceptual framework for environmental analysis of social issues and evaluation of business response patterns. Acad Manag Rev 4(1):63–74. https://doi.org/10.5465/amr.1979.4289184

Seuring S, Müller M (2008) From a literature review to a conceptual framework for sustainable supply chain management. J Cleaner Product 16(15):1699–1710. https://doi.org/10.1016/j.jclepro.2008.04.020

Sharfman M, Ellington RT, Meo M (1997) The next step in becoming 'green': life-cycle oriented environmental management. Busin Horizons 40(3):13–22. https://doi.org/10.1016/S0007-6813(97)90048-3

Shrivastava P (1995) The role of corporations in achieving ecological sustainability. Acad Manag Rev 20(4):936–960. https://doi.org/10.5465/amr.1995.9512280026

Synnestvedt T, Schaltegger S (2002) The link between 'green' and economic success: environmental management as the crucial trigger between environmental and economic performance. J Environ Manag 65(3):339–346. https://doi.org/10.1006/jema

Circular Economy as a Vector for Innovative and Efficient Production and Consumption. Analysis on EU's Indicators

Mihaela Cristina Drăgoi, Jean Vasile Andrei, and Drago Cvijanovic

Abstract Given the numerous problems and concerns related to the linear production-consumption-waste model, employing mechanisms of circular economy (CE) becomes more stringent. This chapter brings into light the main characteristics of the concept of circular economy under the broader spectrum of sustainable development and analyzes the extent to which circular economy indicators and principles are implemented within the European Union, as part of the ambitious targets set out to enhance a more sustainable and greener economy. The assessment of such indicators shows that even though the European Union—taken as whole—is on the right path towards reducing the effects of economic activity on the society and the environment, the situation at Member States' level varies significantly and that sustained efforts and adjustments need to be made by national authorities (as implementation entities) and by the European Commission (as monitoring authority) in order to change production and consumption patterns and to achieve community and international sustainable development and environmental targets.

Keywords Circular economy · European Union · Recycle · Waste · Sustainable development

M. C. Drăgoi (✉)
Bucharest University of Economic Studies, Bucharest, Romania
e-mail: cristina.dragoi@rei.ase.ro

J. V. Andrei (✉)
Petroleum-Gas University of Ploiesti, Ploiesti, Romania

D. Cvijanovic
University of Kragujevac, Kragujevac, Serbia
e-mail: drago.cvijanovic@kg.ac.rs

C. J. Chiappetta Jabbour and S. A. R. Khan (eds.), *Sustainable Production and Consumption Systems*, Industrial Ecology,
https://doi.org/10.1007/978-981-16-4760-4_12

1 Introduction

Since the first Industrial Revolution, modern society has undergone a rapid development in all aspects of life, with machineries—and ulteriorly robots and artificial intelligence (AI)—having contributed significantly to this fast-paced development. Distances have been considerably shortened, working, and living opportunities have changed, the living areas have expanded, the global and regional economy has shifted towards new fields of activity and outputs. There are numerous examples regarding how the world and the society have changed based on the advancements given by the steam engine at first, then by the mass production phase and progress of science, the rise of the digital era all the way through cloud, mobile connections, the Internet of things (IoT) and AI (Britannica 2021; Trailhead 2020). Subsequently, modern citizens have been provided with opportunities and benefits which have become part of the ordinary daily routines and that were purely fantasies or considered unimaginable a few centuries ago.

There is, however, a price tag attached to this economic, technological, and social expansion and the dimensions of this price tag have been widely debated within the press, the public and the academia. At just a brief analysis on this matter, available studies and literature worldwide agree that some of the main shortcomings of the overall development of the society refer—and are not limited—to:

- the increase in the level of pollution;
- the rising quantities of waste deriving from both production and consumption;
- environmental concerns related to global warming's effects and altering natural habitats and resource availability;
- overpopulation in certain areas around the world, especially in urban and peri-urban zones, simultaneously with depopulation in numerous rural areas, leading to less resilience and adaptability;
- accentuation of economic and social imbalances among regions, especially in terms of income, access to public services, working and living opportunities etc.

Taking these aspects into consideration, while human activity and policies were all about finding new means of expansion, in recent decades the society had to shift direction in order to tackle the side-effects of development. Based on all alarm signals regarding the resources, the environment, the effects on human and animal health and to life on earth entirely, current decisions and policies must try to find new means of reversing and reducing damages.

Under these circumstances, there are several answers or feasible solutions for the issues above, meant to reshape the current patterns of production and consumption, of energy and resources use and of civil awareness overall; one of these new approaches involves reducing waste within the context of circular economy. As one of the six priorities of the European Commission for 2024, generally known as the European Green Deal, refers to enhancing the efficiency of resources' use, restoring biodiversity, and reducing pollution, one of the main focuses at European level for the upcoming timeframes is centered towards achieving a cleaner, circular economic activity in order to provide economic and social well-being for the European citizens decoupled from the unwanted environmental effects.

The current chapter addresses the concept of circular economy, starting from various definitions and implications thereof and by putting it in the context of the EU measure, actions, and indicators. The chapter examines the main circular economy indicators as set out by the European Commission and Member States' performance considering each of them, pointing out the frail aspects for both least performing countries and of the EU in its entire when addressing circularity concerns. All in all, in accordance with global sustainable development goals and with the EU's targets, better implementation of circular economy principles is seen as achievable means of counteracting resource scarcity, pollution, demand and supply needs and health and environmental issues, while providing competitive advantage for the EU in international trade.

2 The Concept of Circular Economy. A Brief Literature Review

The concept of *circular economy* has been widely spreading within the academic literature and among organizational studies. Although there are various definitions attributed to this concept and its applicability is still controversial according to the field of implementation, available literature provides several studies on the definitions of circular economy. One of the controversies on this concept derives from the very first step, namely the attempt of defining it in a widely accepted manner. It turns out that various stakeholders in various regions of the world have given different definitions for circular economy. As a consequence, available literature provides numerous interpretations for this concept.

Kirchherr et al. (2017) have identified and catalogued 114 definitions of circular economy, coded on 17 dimensions, finding that the circular economy is most frequently represented as a combination of reduce, re-use and recycle activities, with the core aim of economic prosperity, followed by environmental quality. In their paper, Geissdoerfer et al. (2017) discussed the concepts of *circular economy* and *sustainability* as they both refer to concerns and challenges related to the current state of technology, industrial production, consumption, and environment; by employing the snowballing process, they were able to identify within peer-reviewed English-written indexed papers the main similarities and differences between the two concepts and to point out that the circular economy is viewed as a condition for sustainability, a beneficial relation, or a trade-off.

In sustaining the idea of *closing loops* in industrial ecosystems, Stahel (2016) emphasizes that CE represents a means of turning goods that are at the end of their service life into resources for others, thus minimizing waste. According to the author, the economic logic would be totally different by replacing production with sufficiency—re-using what can be re-used, recycling what cannot be re-used, repairing what is broken, remanufacturing what cannot be repaired.

On the other hand, even though the redesign of processes and material cycling contributes to achieving more sustainable business models, Murray et al. (2017) have identified a scanty social dimension of circular economy; therefore they propose a new definition of the concept as "an economic model wherein planning, resourcing, procurement, production and reprocessing are designed and managed, as both process and output, to maximize ecosystem functioning and human well-being". The study conducted by Korhonen et al. (2018) stresses that achieving progress in sustainable development through the usage of circular economy is possible if both the business community and policy-making community are attracted to sustainability effort.

The same idea also derives from the study conducted by the Ellen MacArthur Foundation, in order to answer some of the European Union's questions regarding the threats deriving from the extensive use of the linear production-consumption-waste model upon the society, the natural resources and the business environment alike. The study emphasizes the need to deepen and accelerate the shift towards a restorative model, which proves beneficial for businesses, economies and society; this shift should be implemented by employing technology, by changing consumer behavior or by raising large-scale awareness as well, all with the firm support of policy-makers and stakeholders also (Ellen MacArthur Foundation 2013). The success stories of circular business models presented in the study enabled launching several consequent objectives and measures within the European Union, from the Europe 2020 strategy all the way to the Green Deal.

As a matter of fact, the sole definition of the CE concept supported by the European Commission refers to maintaining "the value of products, materials and resources for as long as possible by returning them into the product cycle at the end of their use, while minimizing the generation of waste" (Eurostat 2021t). Subsequently, based on the rising prices of resources, supply disruptions, rising competition, declining demand in certain sectors put in balance with excessive waste generation, environmental constraints and health issues, circularity becomes a potential answer to a worldwide range of problems (Adams et al. 2017; Ellen MacArthur Foundation 2013; Geng et al. 2019).

Even though there is no uniquely accepted definition of the concept, circular economy is undoubtfully linked to sustainable development, and apart from the uncontestable economic, social, and environmental benefits, it arises from available literature as a source of unexplored potential competitive advantage.

3 Circular Economy in the European Union—Main Frame

During the past decade, the European Union has become supplementary ambitious in achieving a more sustainable development. Under the umbrella of the United Nations' 17 Sustainable Development Goals (SDGs) which are aiming to improve or eradicate some of the world's major issues—such as hunger, poverty, lack of education, access to clean water and energy, living conditions, environment protection etc. –, the action steps of the European Commission have set the EU on the path of a more efficient greener economy. Under the particular auspices of SDG 12 “Responsible consumption and production”, a new light is shed upon the need of implementing circularity principles in the economy.

Considering these circumstances, the Circular Economy Package was adopted by the European Commission in 2015 along with An EU action plan for the circular economy, aimed at closing the loop between production and consumption focusing on waste management, henceforth tying the circular economy closely to key EU priorities, including jobs and growth, the investment agenda, climate and energy, the social agenda and industrial innovation, and with global efforts on sustainable development (European Commission 2015).

Because there is still a lot to be done in terms of circularity, while also acknowledging efforts made so far, in 2020 the European Commission adopted a new Circular Economy Action Plan—as one of the main pillars of the European Green Deal and of the entire concept of sustainable development of the EU. The measures proposed means of procuring resources for the EU economy without further impacting the environment but with better use of already existing materials through recycling and re-use, all under the comprehensive idea of “less waste, more value” (European Commission 2020).

In order to enact progress in achieving its sustainability objectives, the European Union monitors the circularity of the economy employing a set of indicators as depicted in (Table 1).

The indicators comprised in the first thematic area are of utmost relevance in streamlining production and consumption. Figure 1 presents the percentages of the European Union’s self-sufficiency in extracting raw materials, 100% implicitly signifying that EU is totally self-sufficient. The only raw material category where the EU is almost self-sufficient is Limestone—with more than 95% being extracted within the EU since 2013 –, which is preponderantly used as a building material as a component of concrete, but also utilized in the chemical industry or for interior designing, followed by Copper—more than 55% extracted within the EU ever since 2011—serving as a key-component in the construction sector and power generation and transmission (especially electrical wiring), hence essentially contributing to the production of machinery in general. However, when looking at the other raw materials extraction, the percentages of EU’s self-sufficiency are sensibly lower, even though such materials are important in various industries and activity domains—for example, Cobalt is used all the way from gas turbines to rechargeable batteries, as an essential asset in developing sustainable energy; Aluminium is used for constructing efficiently sealed sky-scrapers, for a wide range of home or industrial appliances, but more importantly for naval and air craft industry, as part of transportation and space knowledge development; Fluorspar and Iron ore are used for the production of Aluminium and steel, while Lithium’s most common uses worldwide are in the production of long-life batteries and heat-resistant glass (CRM Alliance 2021).

There is however a longer list of raw materials which are not available for the EU’s extracting process, but on which current industries depend profoundly, such as Gallium which is the key-element in the semiconductor industry, Germanium which is used in electronics and solar appliances or Indium—an essential element of the

Table 1 European Union's set of indicators for circular economy

Thematic area	Indicator	Sub-indicator
Production and consumption	Self-sufficiency of raw materials for production in the EU	
	Green public procurement (as an indicator for financing aspects)	
	Waste generation (as an indicator for consumption aspects)	Generation of municipal waste per capita
		Total waste generation (excluding mineral) per GDP unit
		Total waste generation (excluding mineral) related to domestic material consumption
	Food waste	
Waste management	Recycling rates (the share of waste which is recycled)	Recycling rate of municipal waste (excluding mineral)
		Recycling rate of all waste (excluding mineral)
	Specific waste streams (packaging waste, biowaste, e-waste)	Recycling rate of overall packaging waste
		Recycling rate of plastic packaging
		Recycling rate of wood packaging
		Recycling rate of electrical and electronic equipment
		Recycled biowaste per capita
		Recovery rate of construction and demolition waste
Secondary raw materials	Contribution of recycled materials to raw materials demand	Secondary raw materials' share of overall material demand—for specific materials
		Secondary raw materials' share of overall material demand—for the whole economy
	Trade of recyclable raw materials between the EU Member States and with the rest of the world	
Competitiveness and innovation	Private investments, jobs and gross value added	Private investments in the circular economy sectors

(continued)

Table 1 (continued)

Thematic area	Indicator	Sub-indicator
		Number of persons employed in the circular economy sectors
		Gross value added in the circular economy sectors
	Number of patents related to recycling and waste management	

Source Eurostat (2021a)

light emitting diodes (LEDs) and laser diodes (CRM Alliance 2021; ESM Foundation 2020d); all these materials either do not have substitutes for their main utilizations, or using their substitute materials results in diminishing the final product's performance.

Therefore, based on the numbers provided by Fig. 1, it is straightforwardly noticeable that there is a high import dependency of the EU for numerous raw materials. According to the ESM Foundation (ESM Foundation 2020a, b, c, d, e, f), one of the main suppliers of the world are: China (for Indium, Germanium, Neodymium, Dysprosium, Europium, Yttrium, Graphite), Congo (for Cobalt) and Australia (for Lithium), followed in a significantly smaller proportion by USA, Russia, Brazil, Canada, Chile, France, Argentina. Additionally, some of these materials are not found in nature, but are extracted when processing other materials (steel or magnets

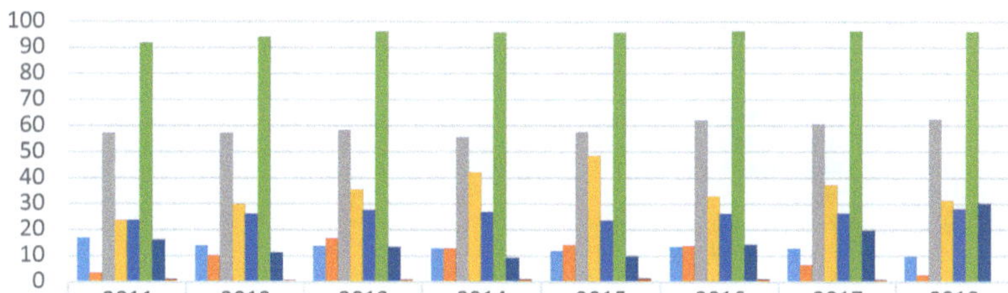

	2011	2012	2013	2014	2015	2016	2017	2018
Aluminium	16.8	14	13.8	12.7	11.9	13.5	12.8	9.8
Cobalt	3.5	10.1	16.6	12.7	14.1	13.8	6.6	2.6
Copper	57.3	57.3	58.4	55.6	57.5	62.1	60.7	62.4
Fluorspar	23.7	29.8	35.5	42.2	48.5	33	37.3	31.3
Iron	23.7	26	27.4	26.7	23.4	26	26.2	28.2
Limestone	91.7	94.1	96.2	95.6	95.6	96.2	96.3	96.2
Lithium	16.1	11.2	13.3	9.1	9.7	14.1	19.6	30.1
Natural graphite	0.8	0.5	0.7	0.8	1.2	0.8	0.7	0.3

Fig. 1 EU[1]'s self-sufficiency for raw materials (%). *Source* authors' development based on Eurostat (2021e)

[1] This chapter considers EU-28 as a reference, thus including United Kingdom until 2020.

for example), consequently recycling used products and materials provides an alternative of procuring such rare raw materials. For the European Union's case, recycling becomes a handy method to increase the supply of raw materials for essential industries, while also reducing the import dependency from other countries, and thus becoming more resilient to worldwide raw materials supply shortages or international negotiations using raw materials as levers. Furthermore, such raw materials are crucial for the development of a European greener economy and recycling boosts a virtuous circle of further enhancement of resource utilization.

Figure 2 emphasizes the extent to which, for a certain raw material, how much of its input into the production system comes from recycling end-of-life products (Eurostat 2021d). the higher the percentage, the lesser need of using primary raw material, which translates into a reduced pressure of extracting natural resources or need of imports, while also ensuring EU's security of supply. The EU's Circular Economy targets also refer to increasing the capability of the Member States to inject back into the economy as many recycled materials as possible, as secondary raw materials, which are beneficial for EU's production, but which also represent a

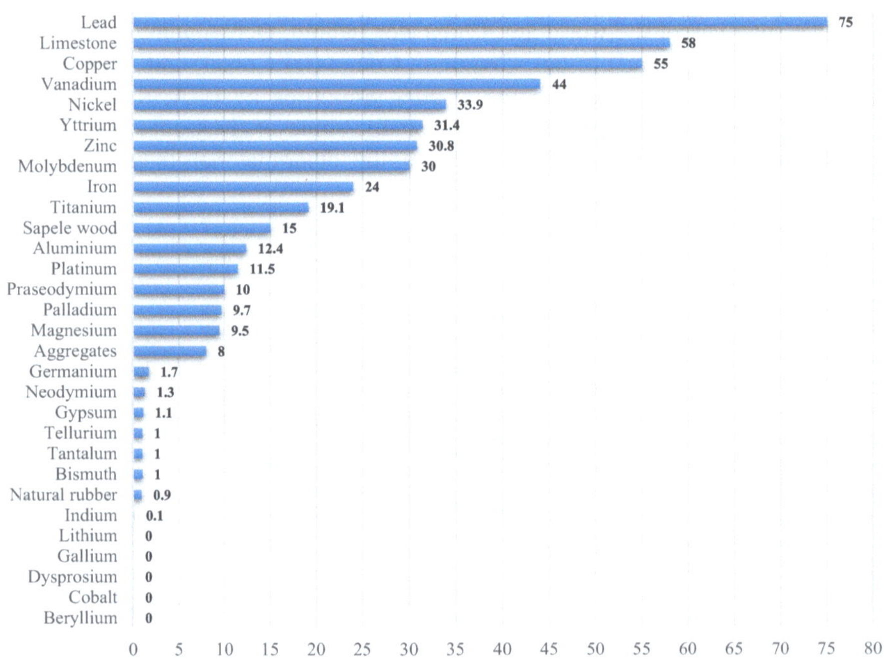

Fig. 2 Contribution of recycled materials to raw materials demand in the EU (2016, %).[2] *Source* authors' development based on Eurostat (2021d)

[2] The term "aggregates" as accepted by the European Commission stands for crushed rock, sands (except silica), pebbles and bituminous additives.

subject of international trade, just like any other extracted primary raw resource, but without the environmental pressure that these imply.

In terms of waste, there is a need for a broader analysis of both waste generation and waste management in the European Union. The generation of municipal waste per capita indicator provides a dimension of the waste collected by municipal authorities, consisting mainly of the waste generated by households, offices, and public institutions, which is then disposed of through the waste management system.

As seen in Table 1, this is one sub-indicator of the waste generation indicator shaping consumption and its evolution should confirm a reduction of the municipal waste generated, underlying a shift in the citizens' patterns of consumption. However, as Fig. 3 emphasizes, there is an ununiform trend among the Member States of the EU. Although, as a whole, the EU-28 has shown a slight decrease in municipal waste generation in 2019 compared to the year 2000, the situation is reversed in the case of certain countries such as: Austria, Croatia, Cyprus, Czechia, Denmark, Finland, France, Greece, Latvia, Lithuania, Luxembourg, Malta, Poland, Portugal, Slovakia, and Sweden, with severe increases in Croatia, Denmark, Luxembourg, Malta, or Slovakia. The most noteworthy reductions in the number of kg. per capita of municipal waste generation are found in Bulgaria, Spain, and the UK, although a decreasing trend can also be seen in Belgium, Estonia, Germany, Hungary, Netherlands, and Romania.

The second sub-indicator among Member States is depicted in Fig. 4, reflecting the waste intensity of the economy. As it covers all economic sectors and households, while excluding major mineral waste, it provides a certain level of comparability among countries, since in some of them major mineral waste comes from extensive construction and mining activities. However, Eurostat statistics take into account the differences in waste classifications among Member States, which may lead to partially non-comparable values. Under these circumstances, the high values shown for Estonia in Fig. 4 can be explained from including the waste generated by energy production into reporting methodology. Therefore, cautious interpretation of the values in this figure shows an overall tendency of reducing the generation of waste when producing wealth (GDP) in almost all Member States; exception to the rule are Denmark, Germany and Italy. Nevertheless, there are several countries in the EU which account for a large share of waste generation related to economic activity, namely Bulgaria, Estonia, Poland, and Romania in comparison to the EU-28 average.

Another dimension of the waste generation indicator unfolds by analyzing the generation of waste (again without considering major mineral wastes) related to domestic material consumption (DMC), specifically to the materials actually used within the national economies. Since both waste and DMC are measured in tons, the ratio between them should be as small as possible in order to prove efficiency. Again, the image determined in Fig. 5 shows that the EU's evolution from 2018 compared to 2004 is uneven among Member States. Even at EU level, the ratio has increased slightly over the past decade, mainly because of the ratio determined in the New Member States. However, during the analyzed timeframe, only half of the Member

States of EU-28 considered (14) have accounted for a decrease of the ratio between waste generation and DMC.

Disaggregation of indicators related to waste generation is understandable, but also raises some alarm signals within EU's countries; while some countries perform better in terms of municipal waste, they are outperformed by other Member States in terms of waste generation related to GDP or DMC. A closer look at available data shows that the EU-13 Member States have joined with a certain gap compared to EU-15 but have accounted for some of the major reductions of waste generation during the past decade.

At EU level, the recycling rates of municipal waste have increased constantly between the years 2000–2019. However, when looking at each country's performance over the past two decades again variations in performance occur (Fig. 6). Countries such as Croatia, Latvia, Lithuania, and Romania did not report any recycling in 2000—according to existing data, in 2019 the recycling rates improved significantly for up to over 30% in Croatia, over 40% in Latvia and almost 50% in Lithuania. On the other hand, Romania registers a modest recycling rate of 11.5%, second to last among all Member States, before the rate of Malta of 8.9%. Although the trend of recycling should be ascending, there is a reverse situation from one period to another in Austria, Belgium, Malta, Romania, and Sweden. As derived from previous information presented in Fig. 3, this reverse tendency is confirmed in the case of Austria, Malta, and Sweden, which not only account for a smaller recycling rate in 2019 compared to 2000, but also generate more municipal waste during the analyzed period. Under these circumstances, the accentuated process of urbanization during the last decades puts an additional pressure on municipalities, as many of the EU's countries are facing a trend of population mobility from rural to urban areas, hence generating supplementary waste, which subsequently may lead to a reduction of the recycling rates and recycling capability of the area.

However, binding legislation, such as the Waste Framework Directive 2008/98/EC (Official Journal of the European Union 2008), has set out a target of 50% of municipal waste recycled by 2020, which, based on 2019 data, was only met in six countries, namely Germany, Slovenia, Austria, the Netherlands, Belgium, Denmark and Italy. Although progress was made in almost all Member States, 13 of them placed below the 40% threshold.

The aforementioned 2015 Circular Economy Action Plan adopted by the European Commission set out more ambitious new targets for 2025–55% of municipal waste recycled and prepared for re-use, for 2030–60% and for 2035–65% (Official Journal of the European Union 2018a). Considering the actual data and the observed slow-paced progress, these targets seem feasible for a limited number of Member States.

The recycling rate of all waste (as previously, excluding major mineral waste) varies significantly across Member States (as observed from Fig. 7). Greece and Latvia did not report any data, and for the latest available year (namely 2016), the discrepancy between countries ranges from an 80% recycling rate in Slovenia to as little as 10% in Estonia. The EU-28 average is situated at 57%, hence 16 countries register a lower recycling rate. In terms of improving the circularity of the economy, returning the generated waste in the economic activities enables better use

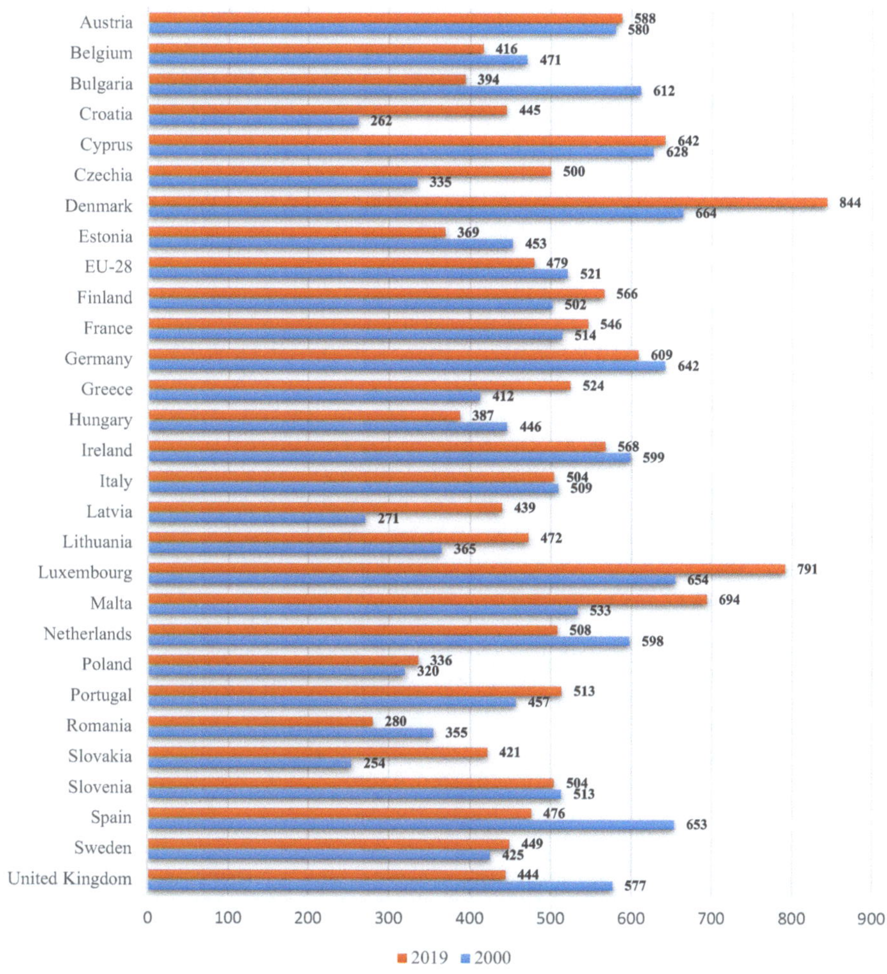

Fig. 3 Generation of municipal waste in the EU (kg. per capita). *Source* authors' development based on Eurostat (2021f)

of resources and reduces losses. Figure 7 emphasizes the striking need of further investments for recycling facilities, as an increased recycling rate would lessen the burden on existing resources.

Another concern for production and consumption processes which is widely taken into consideration, both globally and at EU level, derives from packaging. Packages are highly needed irrespective of the type of product—either raw or processed—in order to protect, handle and deliver products from producers to consumers. The incorrect disposal of packaging has led to numerous effects, especially on the environment, plants and animals' natural habitats and on human health; landfilling and incineration (particularly for plastic packages) have proven to be highly polluting.

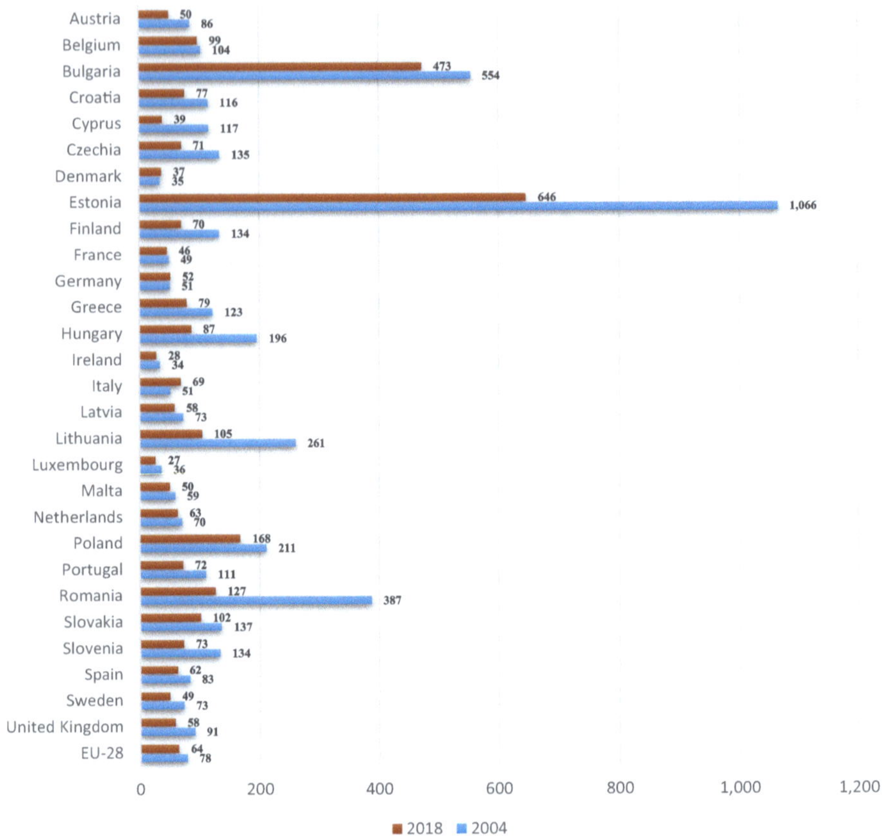

Fig. 4 Generation of waste excluding major mineral wastes per GDP unit. *Source* authors' development based on Eurostat (2021h)

In order to increase the re-usage of packaging, the European Commission has divided packages into 5 different categories—namely wood, paper and cardboard, plastic, glass and metallic—and monitors the recycling action of Member States for all these categories. According to Eurostat, in 2018, the packaging waste generated was positioned at 174 kg. per inhabitant in the EU, with the largest share formed by paper and cardboard (more than 40% of all packaging waste), followed by plastic and glass (around 19% each) (Eurostat 2020).

The Directive 94/62/EC on packaging and packaging waste set out a packaging recycling target of 55% for the year 2008 (Official Journal of the European Union 1994), which, as observed in Fig. 8 was not yet achieved a decade later by all Member States—Hungary and Malta were still behind achieving the 55% threshold in 2018, according to available data. The same directive outlined specific targets for every type of packaging waste: minimum rates of 60% for glass and paper and cardboard, 50% for metals, 22.5% for plastics and 15% for wood, respectively. In 2018, the situation depicted in Fig. 8 showed that for paper and cardboard and plastic packaging the target was met in all Member States. The glass packaging recycling target was not met by Cyprus, Greece, Hungary, Malta, and Portugal, while the metallic packaging recycling target was not met by Croatia, Malta, and Portugal. In the case of wood packaging, the target was not met by Croatia, Cyprus, and Malta, but the recycling rates among Member States vary as significantly as from 90.6% in Belgium to as little as 1.1% in Malta. The percentages higher than 100 are explained by storage amenities, which allow subsequent recycling in the following years.

As part of the EU Action Plan for the Circular Economy, the importance of further enhanced progress in recycling used packaging is stressed out by new targets of 65% recycling rate to be achieved by 2025 and 70% by 2030 (Official Journal of the

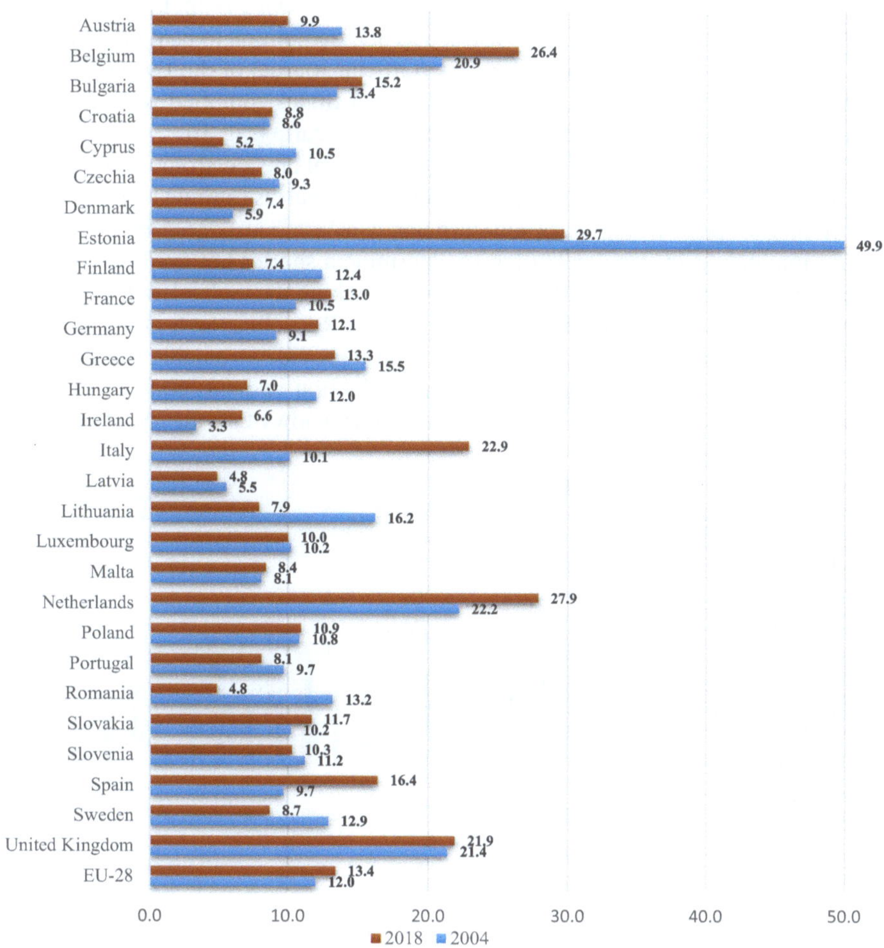

Fig. 5 Generation of waste excluding major mineral wastes per DMC (%). *Source* authors' development based on Eurostat (2021g)

European Union 2018b). There is a promising encouragement, since some Member States have already reached and exceeded the 70% target—namely Belgium, Cyprus (although deficient in recycling wood and glass packaging), Finland, Luxembourg, Netherlands, Slovenia and Sweden; however, the disparities among Member States prove that there is a long way to go in order to enact a unified circular economy in the EU, especially given the EU's specific targets of 55% on recycling plastic packaging by 2025 and 75% on preparation for re-use and recycling of wood packaging by 2030 (Eurostat 2021p).

Another vital recycling opportunity derives from constructions and demolitions, as many of the materials employed in this field are recyclable or re-usable and represent a valuable source for raw materials as well. As construction and demolition stand for one of the major sources of waste in the European Union, the Waste Framework Directive 2008/98/EC has set a mandatory target by 2020 which stipulates a minimum of 70% of the waste generated by construction and demolition to be prepared for re-use, recycling and other material recovery (Official Journal of the European Union 2008). In 2018, the majority of the EU's Member States complied with the target, the only exceptions being Cyprus, Slovakia, and Bulgaria (Fig. 9).

The indicator of e-waste or WEEE (waste electrical and electronic equipment) is another source of raw materials (especially rare ones such as Lithium and Indium) on which the European Union lacks self-sufficiency as previously seen in this section. Innovation and developing technologies determine a shorter life cycle for products such as computers, TV-sets and mobile phones, making them obsolete faster than in previous decades, hence increasing consumerism. Recycling such products would reduce furthermore the waste footprint and would provide a valuable source of rare raw materials, while reducing the costs of production and subsequently the final retailing prices. Compared to 2010, data for 2018 shows that all Member States,

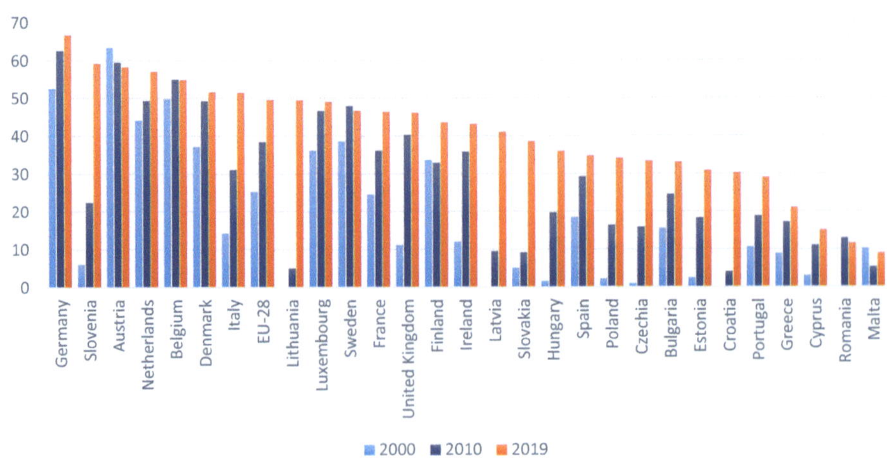

Fig. 6 Recycling rate of municipal waste in the European Union (%). *Source* authors' development based on Eurostat (2021o)

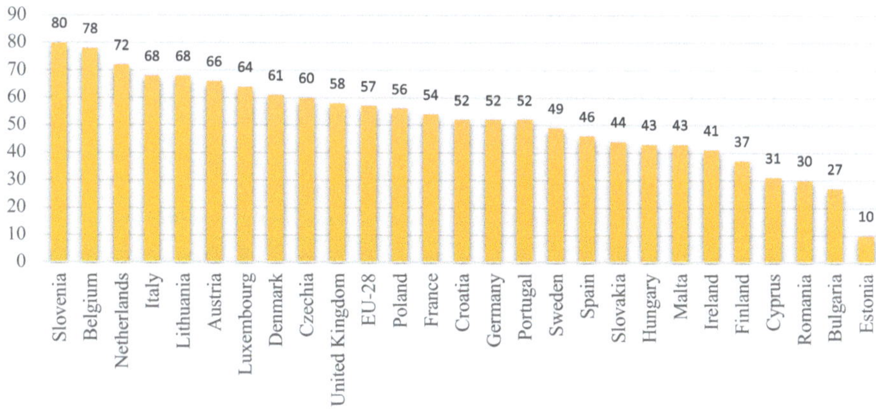

Fig. 7 Recycling rate in the EU of all waste excluding major mineral waste (%, 2016). *Source* authors' development based on Eurostat (2021m)

except to a higher extent Sweden and a lesser extent Germany, have achieved progress in increasing the recycling rate of e-waste (Fig. 10). According to the WEEE Directive 2012/19/EU, the target recycling rates vary in line with the type of electrical and electronic equipment, ranging from a minimum of 55% in the case of small IT and telecommunication equipment, toys and sports equipment or medical devices to up to 80% for large household appliances, automatic dispensers or lamps (Official Journal of the European Union 2012).

As previously underlined, the circular economy concept leads to improving the means of recycling and re-using waste derived from economic activities and living patterns. In this sense, the circular material use rate measures the actual share of materials recovered and placed back into the economy; the circularity rate as shown in Fig. 11 highlights the extent to which recycled materials are used for the economic activity as a substitute for extracting primary raw materials; subsequently, a higher circularity rate translates into the reduction of the environmental impact of extracting primary materials.

As observed, the circularity rate itself is lower than recycling rates analyzed before. An explanation for this situation is that there are various materials which cannot be recycled and re-used, as for example the biomass consumed as fodder for animals or the fossil fuels as petroleum or gas burned in order to produce energy. Additionally, the circularity rate itself is calculated as the ratio of the circular use of materials to the overall material use, which, at its turn, comprises the aggregate DMC and the circular use of materials. The latter is defined by Eurostat as the amount of waste recycled in domestic recovery plants minus imported waste destined for recovery plus exported waste destined for recovery abroad (Eurostat 2021b). Under this light, in countries where the recycling rate is high, the circularity rate is also higher. Furthermore, DMC is estimated as the quantity of raw materials extracted within a country, plus all material imports minus exports. If the DMC is low, then the circularity rate turns up higher, which can be explained either by low domestic material consumption of

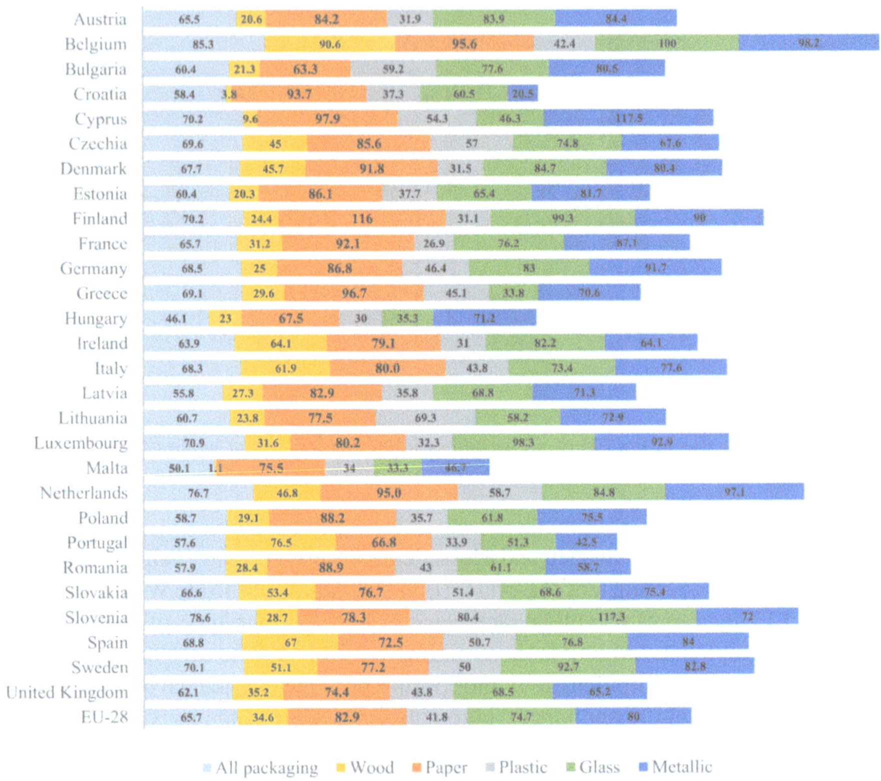

Fig. 8 Recycling rate of packaging in the EU, by type of package, in 2018 (% of all packaging waste). *Source* authors' development based on Eurostat (2021p)

a country, namely low local extractions of materials, or by low material imports or high material exports.

A noteworthy dimension is added by the amount of waste imported to be recycled locally or exported to be recycled elsewhere, as they impact the bottom-line ratio of the circular material use. Countries such as Netherlands and Belgium are net exporters of waste hence the greater circularity rate they account for (Fig. 12). The classes of waste taken into account refer to: plastic; paper and cardboard; precious metal; iron and steel; copper, aluminium and nickel and shipments are classified as between Member States (intra-EU), and with third countries (extra-EU). The countries which are net importers of waste and account for a more reduced recycling rate are positioned lower in the circularity rate ranking. However, caution is again needed when interpreting strictly from a mathematical point of view the matter of import and export of waste, since improved waste recycling facilities determine an increase in the possibility of producing secondary material for the economy, thus diminishing the pressure on extracting or importing primary raw material; in this sense, import of waste, although affects the resulting calculations of circularity rates

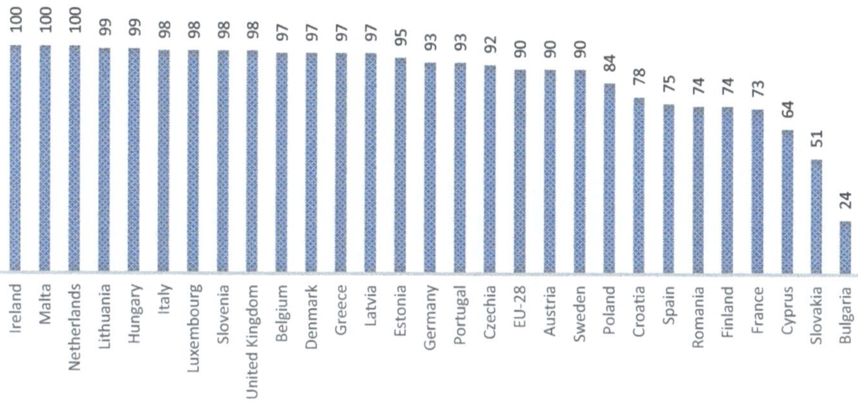

Fig. 9 Recovery rate of construction and demolition waste in the EU (2018, %). *Source* authors' development based on Eurostat (2021l)

does not necessarily imply a disadvantage if waste is treated and recycled accordingly so as to put it again at use in production.

A closer look at EU's DMC shows that almost half of the domestic material consumption is attributed to non-metallic minerals, which account for a circularity rate of 15.8% (Fig. 13). Metal ores register only 5%, but they account for the highest circularity rate, hence being re-used within the economy at a larger scale. Fossil energy materials are the least recyclable materials among all four categories while also highly pollutant for the environment, hence the ambitious objectives of the EU of lowering the use of fossil energy materials for a greener European economy. Eurostat data emphasizes that the consumption patterns vary across Member States, most significantly in the case of non-metallic minerals ranging from around 2 tons per capita to more than 21 tons per capita, variations explained by the country's endowment with natural resources, population density, construction activities and dimension of the infrastructure (Eurostat 2021s). certain patterns of specialization result in higher consumption as well, for instance in the case of livestock or timber production, consumption of biomass is higher.

Given that DMC comprises domestic extraction and imports of raw materials and that non-metallic minerals are mostly used within the EU's DMC, the aforementioned variability in consumption patterns is also noticeable when analyzing the material import dependency, ranging from 93% import dependable in Luxembourg to only 2% in Romania in the case of non-metallic minerals (Fig. 14). The necessity of importing materials in order to sustain national economy varies significantly across material categories and across Member States; higher rates (close or equal to 100%) signify that the countries are almost or entirely relying on imports to ensure their necessary material flows for production. Thus, a new light is shed upon the relevance of reinforcing the recycling capabilities of the EU's Member States so as to increase the share of secondary raw materials usage within economic activity.

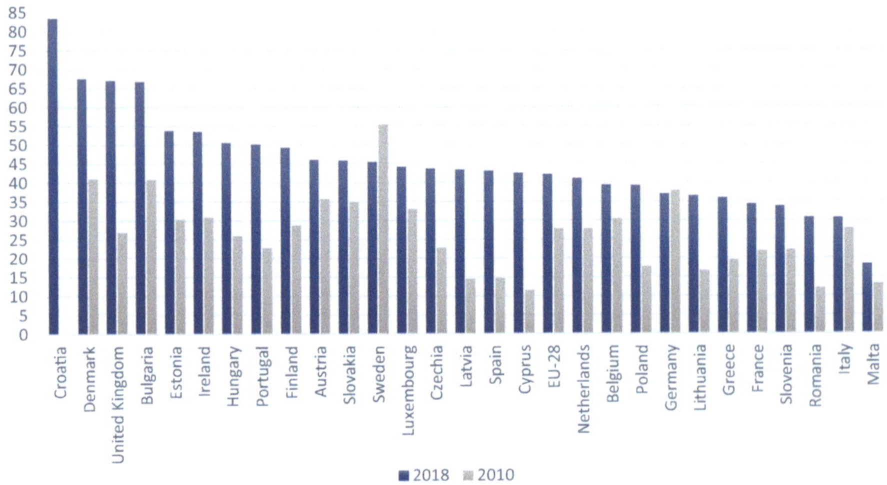

Fig. 10 Recycling of e-waste in the EU (%). *Source* authors' development based on Eurostat (2021n)

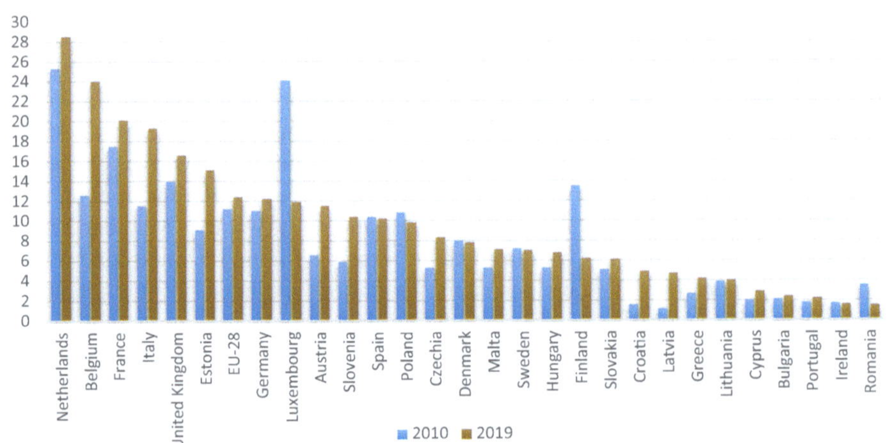

Fig. 11 Circular material use rate in the EU (% of total material use). *Source* authors' development based on Eurostat (2021b)

However, when running DMC against GDP, namely as the ratio between GDP (in purchasing power standard—PPS, for comparability purposes) as in Fig. 15, and the domestic material consumption in order to determine the value of GDP generated for each unit of resources utilized within the economy, we obtain an EU-28 average of 2.33 euro/kg. for 2019. Across the timeframe under analysis (2019 compared to 2010) resource productivity increased, the only exceptions being Malta, Sweden, Hungary, and Romania. Additionally, the gap between resource productivity among

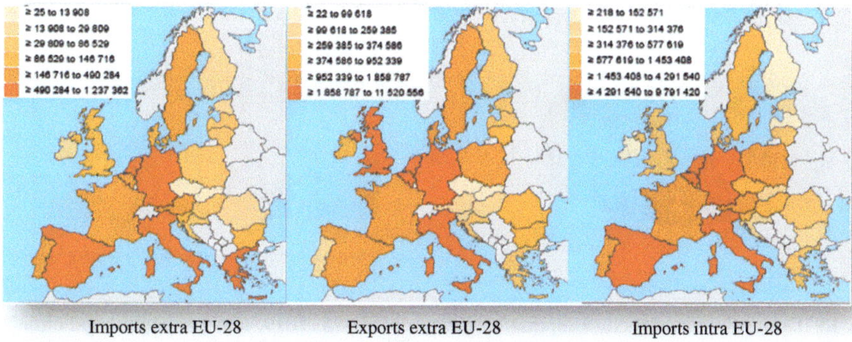

Fig. 12 Trade in recyclable raw materials (2019). *Source* authors' development based on Eurostat (2021r)

Member States has widened with more than 3.5 euro/kg. in Belgium, Luxembourg, Italy, United Kingdom and Netherlands and less than 1 euro/kg. in Estonia, Romania, and Bulgaria. There is an observed correlation between the higher recycling and circularity rates and higher resource productivity, as well as between high DMC and lower resource productivity as deriving from previous figures. Hence, countries with low GDP and higher DMC such as Romania, Estonia, and Bulgaria account for a diminished resource productivity, while good performing countries in terms of circularity account for a higher GDP and lower DMC, namely higher resource productivity, as in the case of Netherlands.

The graph provided by Eurostat (Fig. 16) expresses in a visual manner the extent to which the environment is relieved by the pressure exerted by economic growth. As previously seen, DMC stands for extraction of raw materials, which are then used (along with imported materials) within the economic activity in order to produce economic well-being (i.e., GDP). In order to reduce the impact on the environment, economic growth should be decoupled from material consumption—as defined by the Organization for Economic Co-operation and Development (OECD)—in conjunction with developing recycling processes and increasing resource productivity. According to Fig. 16, the EU Member States divide into three categories related to DMC and GDP growth decoupling. The group situated below the diagonal line—which represents equal change rates of both GDP and DMC—consists of countries whose GDP increased at a faster rate than DMC, therefore attaining a partial decoupling. The group of countries—including EU as a whole—situated below the horizontal line succeeded to achieve absolute decoupling, signifying that domestic material consumption has decreased while GDP has increased during the analyzed period. Only two countries situated above the diagonal line, namely Estonia and especially Romania did not decouple DMC from GDP, indicating that GDP growth is strongly correlated to an increase of domestic material consumption, implying furthermore that there is a strong pressure on domestic extraction or high dependency of import of materials. Again, better performing countries in terms of implementing circular economy conducts are placed on the absolute decoupling side of the graph.

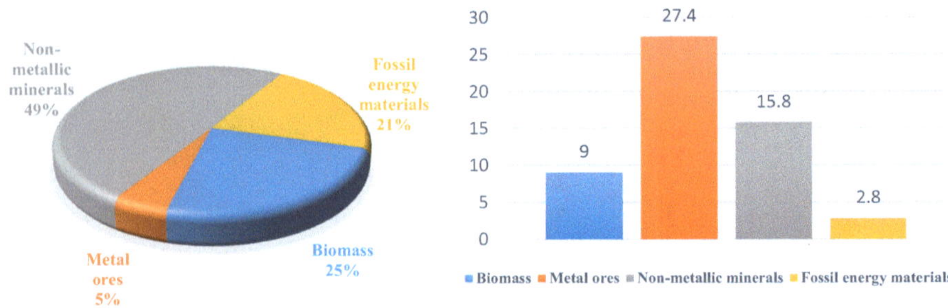

Fig. 13 Domestic material consumption by category (EU-28, 2019, tons per capita) and circular material use rate by material type (EU-28, 2019, %). *Source* authors' development based on Eurostat (2021c, i)

Employing the principles of circular economy encompasses a certain level of innovation, as the linear model of production-consumption-waste has already proven its weaknesses and its effect on the environment. Figures 17 and 18 show that there is more potential to be exploited across Member States and by comparison to other countries of the world.

The most impressive increase in patents related to recycling and secondary raw materials is noticed in the case of China, which accounted for almost 6,500 patents in 2016 compared to only 146 in 2000. If we correlate these absolute numbers with the dimension of the national population, then the figures are more favorable to smaller countries, such as Luxembourg, Finland, or Estonia.

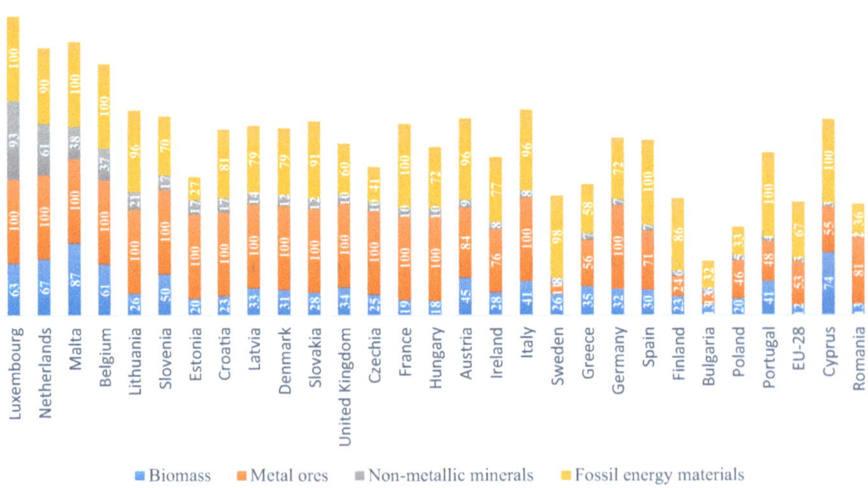

Fig. 14 Material import dependency by category (EU-28, 2019, %). *Source* authors' development based on Eurostat (2021j)

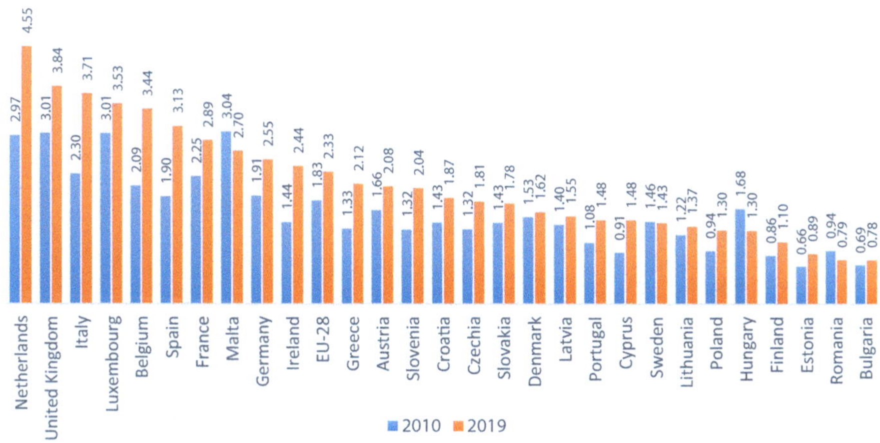

Fig. 15 Resource productivity in the EU (PPS per kg.). *Source* authors' development based on Eurostat (2021q)

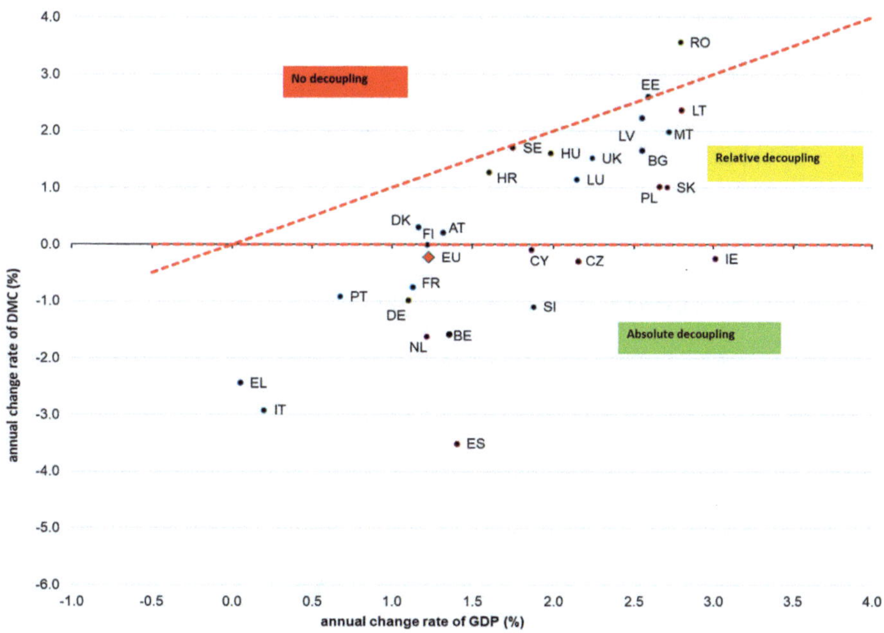

Fig. 16 Changes of DMC and GDP by country (2000–2019, annual exchange rates in %). *Source* Eurostat (2021u)

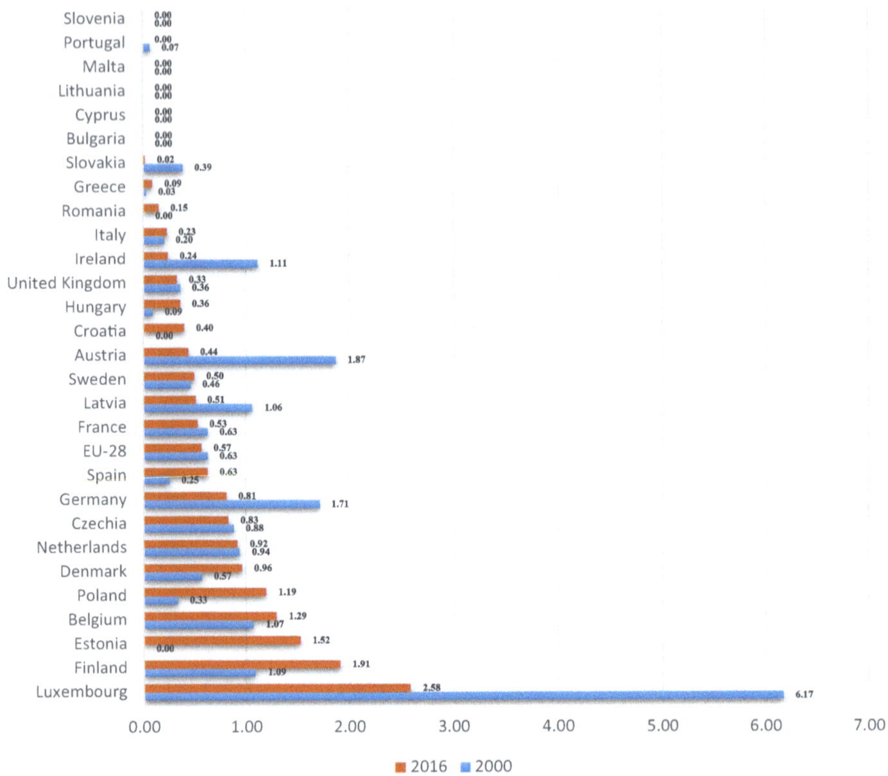

Fig. 17 Patents related to recycling and secondary raw materials in the EU (per million inhabitants). *Source* authors' development based on Eurostat (2021k)

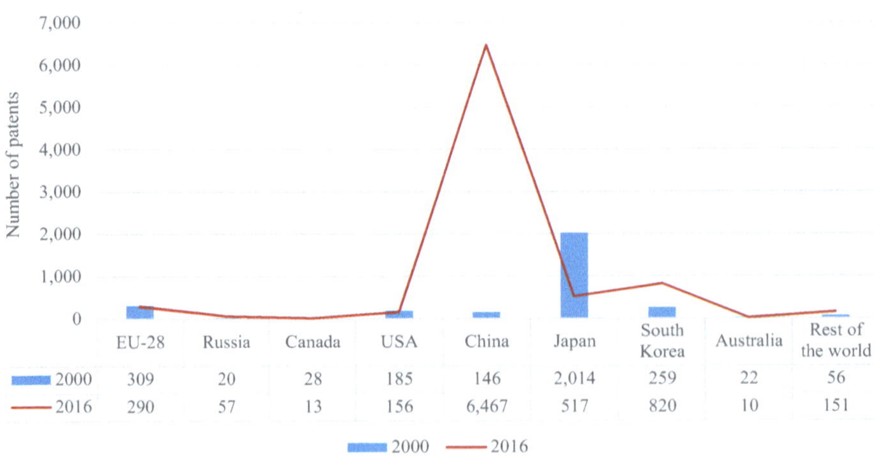

Fig. 18 Number of patents related to recycling and secondary raw materials. *Source* authors' development based on Eurostat (2021k)

4 Discussion and Conclusion

In 2018, the European Union—comprising 28 countries—generated 1,820 kg. per capita of waste (excluding major mineral waste) of which 495 kg. per capita of municipal waste. If left as purely waste, it represents a massive loss of resources and has a great impact on the environment—landfilling implies occupying large surfaces of land causing air, water, and soil pollution, while incineration generates air pollution. The concept of circular economy widely debated in this chapter provides means of transforming such threats into opportunities. Under the circular economy umbrella, waste receives prospects of becoming materials and energy, which can be employed again and again in the economic activity.

The aims of the new circular economy paradigm proposed by the European Commission—following EU's own ambitions, as well as international agreements and protocols—are meant to change the patterns of production of goods and provision of services towards achieving long-term sustainability goals. On EU's path to a cleaner and greener economy, the elements of circularity are promoted as some of the main pillars. Under these circumstances, EU policies and key-actions are meant to ensure either the safe disposal of waste, or better yet to reduce the quantity of generated waste and to reintroduce the unavoidable generated waste back into the economic activity.

As the sustainability objectives of the EU economic activities are closely linked to international desiderates regarding safeguarding existing resources, protecting the environment, improving overall health, and preserving ecosystems along with delivering clean and dependable products and services to customers, the circular economy mechanisms provide the frame to improve resource productivity and to reduce the amount of waste associated to economic activity and private consumption.

Additionally, more efficient consumption is directly related to diminishing food waste. According to Stenmarck et al. (2016), in EU approximately 20% of total food produced is lost or wasted, with more than half of this percent pertaining to households. While there are still more than 30 million people who cannot afford a quality meal every second day (European Commission 2021), diminishing food waste across all stages of the food chain is of utmost importance, also stressed in the Farm to Fork Strategy adopted by the European Commission in May 2020 as part of the Green Deal plan.

In order to achieve the greener economy desiderate, the EU's subsequent conduct puts an additional focus on public procurement as an indicator for the circularity of the European economy, claiming that if public contracts would include more circularity requirements, then public procurement's role in CE would increase significantly and would create a training effect for other economic actors as well—some of the examples of good practices delivered by the European Commission from various Member States provide sufficient incentives for public authorities to enable circularity models in their procurements, which account for approx. 14% of EU GDP (European Commission 2017; Neubauer et al. 2017).

Hence, a key factor in creating a dynamic market for secondary raw materials is raising awareness of producers and consumers as well, in order to create sufficient demand, powered by the use of recycled materials in products and infrastructure. For certain raw materials (such as paper or metal), there is already an increased demand, but for others, it is still developing. The role of both the public and private sectors in creating demand and helping to shape supply chains will be essential. As pointed out by the European Commission (2015), there is an encouraging perspective to the matter, as various industrial and economic actors have already given public commitment to ensuring a certain level of recycled content in products they sell for both sustainability and economic reasons.

A noteworthy dimension is related to the opportunities implied and enhanced by the application of the principles of circular economy. As the European Commission estimates for 2021–2018 show, the number of jobs associated to the circular economy in the EU increased by 5% to reach approx. 4 million ((European Commission 2020); the positive effect of circularity on jobs creation is conditioned by workers acquiring the skills demanded by the green transition and is also sustained by academic research, as the study conducted by Horbach et al. (2015) demonstrates a direct positive effect between circular economy and green jobs employment.

Under these circumstances, performance among Member States should equalize in development, as the discrepancies pointed out throughout the chapter make it difficult to assess an overall performance of the EU in this matter. As previously seen, there are leading countries which overclassify other Member States in terms of efficient usage of resources, recycling and re-using materials, while the latter still have to strengthen the means to mitigate resource scarcity, pollution, or welfare gaps. Countries such as Denmark, Luxembourg and Netherlands generate significant amount of municipal waste for example, compared to countries such as Romania (only 280 kg. per capita of municipal waste, compared to 844 kg. per capita in Denmark), Estonia or Hungary, but also account for a higher recycling rate of that waste. Countries such as Malta are outperformed in terms of both waste generation and waste recycling, hence further investments and rethinking national policies are highly needed.

Although at EU level figures look promising, the deeper analysis performed in this chapter, which highlighted the circular economy indicators at national level for the Member States, has shown the main weak points of delivering a more profound circular economy and reaching sustainable development in a generalized manner throughout the European Union. Infringement procedures initiated by the European Commission regarding defections landfilling and waste incineration against several Member States (again Romania and Malta included, as poor legislative implementation) demonstrate that sustainable progress and public authorities' and private public's awareness are not substantial in all Member States.

Following transposition into national legislation, according to Directives (EU) 2018/851 and (EU) 2018/852, Member States should be constrained to deliver the targets leading towards a more circular economy within the EU, namely:

- to achieve the minimum levels of re-use and recycle of packaging waste;

- to make extended producer responsibility (EPR) schemes mandatory for packaging, electrical and electronic equipment, batteries and accumulators, and end-of-life vehicles—implying that producers take over the financial and/or organizational responsibility for collecting or taking back used goods, as well as sorting and treatment for their recycling;
- to consolidate the implementation of the waste hierarchy—respectively addressing waste in the order of prevention, preparing for re-use, recycling, recovery and ultimately disposal;
- to strengthen requirements related to separate waste collection.

Being synergic with EU's other objectives related to climate and clean energy, circular economy encompasses actions and measure covering all the life cycle of products: from production and consumption to waste management and the market for secondary raw materials and given its impact on generating economic opportunities and social and environmental improvements, such efforts should provide sufficient incentives for stronger and more sustained action by Member States. additionally, smart products design and production processes have demonstrated to enable resources saving and to avoid waste generation. As this implies reshaping business models, circular economy unquestionably leads to the creation of new business opportunities, thus reinventing the economy on the producer side towards being more sustainable and competitive. Also, by providing the means for end-of-life products collection or for extending products lifespan through repairing and spare parts usage, it ensures avoiding premature obsoleteness and a wider, cleaner, and cost-saving product offer for consumers. All aspect considered, circular economy as a vector of sustainable development implies further shifting patterns towards more innovative and efficient ways of producing and consuming.

But, as Stahel (2016) pointed out, circular economy knowledge is concentrated mainly in big industries and scattered among small-medium enterprises (SMEs). As formerly highlighted regarding jobs creation within the circular economy, if such knowledge is brought into academic and vocational training, a bottom-up approach of circular economy would escalate more quickly, as SMEs would have the appropriately trained staff to change business models. An integrated view of implementing circular economy therefore proves useful, implying authorities and regulators as policy and decision-makers and scientists and academia as levers for innovation and knowledge disseminators alike.

References

Adams KT, Osmani M, Thorpe T, Thornback J (2017) Circular economy in construction: current awareness, challenges and enablers. Proc Inst Civil Eng - Waste Resour Manag 170(1):15–24. https://doi.org/10.1680/jwarm.16.00011

CRM Alliance (2021) Critical raw materials. https://www.crmalliance.eu

Britannica (2021) Industrial revolution. https://www.britannica.com/event/Industrial-Revolution

European Commission (2015) Closing the loop – an EU action plan for the circular economy. https://eur-lex.europa.eu/legal-content/EN/TXT/?qid=1453384154337&uri=CELEX:52015DC0614
European Commission (2017) Public procurement for a circular economy. https://ec.europa.eu/environment/gpp/pdf/Public_procurement_circular_economy_brochure.pdf
European Commission (2020) A new circular economy action plan for a cleaner and more competitive Europe. https://eur-lex.europa.eu/legal-content/EN/TXT/?qid=1583933814386&uri=COM:2020:98:FIN
European Commission (2021) Stop food waste. Food safety. https://ec.europa.eu/food/safety/food_waste/stop_en
Ellen MacArthur Foundation (2013) Towards the circular economy. Economic and business rationale for an accelerated transition. https://www.werktrends.nl/app/uploads/2015/06/Rapport_McKinsey-Towards_A_Circular_Economy.pdf
Eurostat (2020) Packaging waste statistics. Statistics explained. https://ec.europa.eu/eurostat/statistics-explained/index.php/Packaging_waste_statistics
Eurostat (2021a) Circular economy. Indicators. https://ec.europa.eu/eurostat/web/circular-economy/indicators
Eurostat (2021b) Data browser – Circular material use rate. https://ec.europa.eu/eurostat/databrowser/view/CEI_SRM030__custom_354994/bookmark/map?lang=en&bookmarkId=c6638243-2f7f-4256-b2fd-6a5159b4b68a
Eurostat (2021c) Data Browser – Circular material use rate by material type. https://ec.europa.eu/eurostat/databrowser/view/env_ac_curm/default/table?lang=en
Eurostat (2021d) Data Browser – Contribution of recycled materials to raw materials demand - end-of-life recycling input rates. https://ec.europa.eu/eurostat/databrowser/view/CEI_SRM010__custom_354964/bookmark/bar?lang=en&bookmarkId=3e951583-9e23-4ecd-ab03-4060eb845b83
Eurostat (2021e) Data browser – EU self-sufficiency for raw materials. https://ec.europa.eu/eurostat/databrowser/view/cei_pc010/default/table?lang=en
Eurostat (2021f) Data browser – Generation of municipal waste per capita. https://ec.europa.eu/eurostat/databrowser/view/CEI_PC031__custom_354618/bookmark/table?lang=en&bookmarkId=ba30dc3d-dfe5-4de2-a445-4cbadb442497
Eurostat (2021g) Data browser – Generation of waste excluding major mineral wastes per domestic material consumption. https://ec.europa.eu/eurostat/databrowser/view/CEI_PC033__custom_354699/bookmark/table?lang=en&bookmarkId=1eaf6e19-68e5-48b0-8d7c-b76ed30d90ef
Eurostat (2021h) Data browser – Generation of waste excluding major mineral wastes per GDP unit. https://ec.europa.eu/eurostat/databrowser/view/CEI_PC032__custom_354647/bookmark/table?lang=en&bookmarkId=6861ca21-d5c8-4fc1-9d66-eabee18da732
Eurostat (2021i) Data browser – Material flow accounts. https://ec.europa.eu/eurostat/databrowser/view/ENV_AC_MFA__custom_819278/default/table?lang=en
Eurostat (2021j) Data browser – Material import dependency. https://ec.europa.eu/eurostat/databrowser/view/env_ac_mid/default/table?lang=en
Eurostat (2021k) Data browser – Patents related to recycling and secondary raw materials. https://ec.europa.eu/eurostat/databrowser/view/CEI_CIE020__custom_355163/bookmark/table?lang=en&bookmarkId=2f9c5749-a486-4dbe-8cd9-974884540f79
Eurostat (2021l) Data browser – Recovery rate of construction and demolition waste. https://ec.europa.eu/eurostat/databrowser/view/CEI_WM040__custom_354944/bookmark/table?lang=en&bookmarkId=7e4febc4-a0fd-444b-aae0-94643539ce0d
Eurostat (2021m) Data browser – Recycling rate of all waste excluding major mineral waste. https://ec.europa.eu/eurostat/databrowser/view/CEI_WM010__custom_354779/bookmark/table?lang=en&bookmarkId=82b41769-ad78-49d9-a470-600e923b1f4f
Eurostat (2021n) Data browser – Recycling rate of e-waste. https://ec.europa.eu/eurostat/databrowser/view/CEI_WM050__custom_354886/bookmark/table?lang=en&bookmarkId=4a9433ab-f0ad-45bd-bcb7-0321289cb2c8

Eurostat. (2021o) Data browser – Recycling rate of municipal waste. https://ec.europa.eu/eurostat/databrowser/view/CEI_WM011__custom_354731/bookmark/table?lang=en&bookmarkId=8dc83e1a-de5d-4d1f-8150-7a8cda664888

Eurostat (2021p) Data browser – Recycling rate of packaging waste by type of packaging. https://ec.europa.eu/eurostat/databrowser/view/CEI_WM020__custom_354828/bookmark/table?lang=en&bookmarkId=d3ad391e-e2c7-4c78-9a0e-542d309b8084

Eurostat (2021q) Data browser – Resource productivity. https://ec.europa.eu/eurostat/databrowser/view/env_ac_rp/default/table?lang=en

Eurostat (2021r) Data browser – Trade in recyclable raw materials. https://ec.europa.eu/eurostat/databrowser/view/CEI_SRM020__custom_355078/bookmark/map?lang=en&bookmarkId=fe4417e4-0764-4a46-8c91-0ed5a9007110

Eurostat (2021s) Material flow accounts and resource productivity. Statistics explained. https://ec.europa.eu/eurostat/statistics-explained/index.php?title=Material_flow_accounts_and_resource_productivity

Eurostat (2021t) Overview - Circular economy. https://ec.europa.eu/eurostat/web/circular-economy

Eurostat (2021u) Resource productivity statistics. Statistics explained. https://ec.europa.eu/eurostat/statistics-explained/index.php/Resource_productivity_statistics

ESM Foundation (2020a) Material of the month: Cobalt. https://www.esmfoundation.org/material-of-the-month-cobalt/

ESM Foundation (2020b) Material of the month: Germanium. https://www.esmfoundation.org/material-of-the-month-germanium/

ESM Foundation (2020c) Material of the month: Graphite. https://www.esmfoundation.org/material-of-the-month-graphite/

ESM Foundation (2020d) Material of the month: Indium. https://www.esmfoundation.org/material-of-the-month-indium/

ESM Foundation (2020e) Material of the month: Lithium. https://www.esmfoundation.org/material-of-the-month-lithium/

ESM Foundation (2020f) Material of the month: Rare Earth Elements. https://www.esmfoundation.org/material-of-the-month-rare-earth-elements/

Geissdoerfer M, Savaget P, Bocken NMP, Hultink EJ (2017) The circular economy – A new sustainability paradigm? J Cleaner Product 143:757–768. Elsevier Ltd. https://doi.org/10.1016/j.jclepro.2016.12.048

Geng Y, Sarkis J, Bleischwitz R (2019) How to globalize the circular economy. Nature 565(7738): 153–155. Nature Research. https://doi.org/10.1038/d41586-019-00017-z

Horbach J, Rennings K, Sommerfeld K (2015) Circular economy and employment. https://www.google.com/search?client=firefox-b-d&q=Centre+for+European+Economic+Research+%28ZEW%29++contact

Kirchherr J, Reike D, Hekkert M (2017) Conceptualizing the circular economy: an analysis of 114 definitions. Resour, Conserv Recycl 127:221–232. https://doi.org/10.1016/j.resconrec.2017.09.005

Korhonen J, Honkasalo A, Seppälä J (2018) Circular economy: the concept and its limitations. Ecol Econ 143:37–46. https://doi.org/10.1016/j.ecolecon.2017.06.041

Murray A, Skene K, Haynes K (2017) The circular economy: an interdisciplinary exploration of the concept and application in a global context. J Bus Ethics 140(3):369–380. https://doi.org/10.1007/s10551-015-2693-2

Neubauer C, Jones M, Montevecchi F, Schreiber H, Tisch A, Walter B (2017) Green public procurement and the EU action plan for the circular economy.

Official Journal of the European Union (1994) European Parliament and Council Directive 94/62/EC of 20 December 1994 on packaging and packaging waste. https://eur-lex.europa.eu/legal-content/EN/TXT/?uri=CELEX:01994L0062-20150526

Official Journal of the European Union (2008) Directive 2008/98/EC of the European Parliament and of the Council of 19 November 2008 on waste and repealing certain Directives. https://eur-lex.europa.eu/legal-content/EN/TXT/?uri=celex%3A32008L0098

Official Journal of the European Union (2012) Directive 2012/19/EU of the European Parliament and of the Council of 4 July 2012 on waste electrical and electronic equipment (WEEE). https://eur-lex.europa.eu/legal-content/EN/TXT/?uri=celex%3A32012L0019

Official Journal of the European Union (2018a) Directive (EU) 2018/851 of the European Parliament and of the Council of 30 May 2018 amending Directive 2008/98/EC on waste. https://eur-lex.europa.eu/legal-content/EN/TXT/PDF/?uri=CELEX:32018L0851&rid=1

Official Journal of the European Union (2018b) Directive (EU) 2018/852 of the European Parliament and of the Council of 30 May 2018 amending Directive 94/62/EC on packaging and packaging waste. https://eur-lex.europa.eu/legal-content/EN/TXT/?qid=1528981579179&uri=CELEX:32018L0852

Stahel WR (2016) The circular economy. Nature 531(7595):435–438. Nature Publishing Group. https://doi.org/10.1038/531435a

Stenmarck Å, Jensen C, Quested T, Moates G (2016) Estimates of European food waste levels.

Trailhead (2020) Meet the three industrial revolutions. https://trailhead.salesforce.com/en/content/learn/modules/learn-about-the-fourth-industrial-revolution/meet-the-three-industrial-revolutions

MIX
Papier aus verantwortungsvollen Quellen
Paper from responsible sources
FSC® C105338

If you have any concerns about our products,
you can contact us on
ProductSafety@springernature.com

In case Publisher is established outside the EU,
the EU authorized representative is:
Springer Nature Customer Service Center GmbH
Europaplatz 3, 69115 Heidelberg, Germany

Printed by Libri Plureos GmbH
in Hamburg, Germany